T0138527

THE POLITICS
OF
PURE SCIENCE

THE POLITICS
OF
PURE SCIENCE

Daniel S. Greenberg

New Edition

With introductory essays by
JOHN MADDOX
and
STEVEN SHAPIN
and a new Afterword
BY THE AUTHOR

THE UNIVERSITY OF CHICAGO PRESS
CHICAGO AND LONDON

DANIEL S. GREENBERG is a syndicated newspaper columnist and visiting scholar in the Department of the History of Science, Medicine, and Technology at Johns Hopkins University. He is founding editor and was publisher of the *Science and Government Report* from 1971 to 1997.

SIR JOHN MADDOX is Editor Emeritus of *Nature*. His books include *Revolution in Biology, The Doomsday Syndrome, Beyond the Energy Crisis,* and *What Remains to Be Discovered: Mapping the Secrets of the Universe, the Origins of Life, and the Future of the Human Race.*

STEVEN SHAPIN is professor of sociology at the University of California, San Diego. His books include *A Social History of Truth: Civility and Science in Seventeenth-Century England, The Scientific Revolution,* and *Science Incarnate: Historical Embodiments of Natural Knowledge* (coedited with Christopher Lawrence), all published by the University of Chicago Press.

The University of Chicago Press, Chicago 60637
The University of Chicago Press, Ltd., London
© 1967, 1999 by Daniel S. Greenberg
© 1999 by The University of Chicago
All rights reserved. Published 1999
Printed in the United States of America
07 06 05 04 03 02 01 00 99   1 2 3 4 5
ISBN: 0-226-30631-3 (cloth)
ISBN: 0-226-30632-1 (paper)

Library of Congress Cataloging-in-Publication Data

Greenberg, Daniel S., 1931-
    The politics of pure science / Daniel S. Greenberg ;
    with introductory essays by John Maddox and Steven Shapin
    and a new afterword by the author.
        p.   cm.
    Includes bibliographical references and index.
    ISBN 0-226-30631-3 (cloth : alk. paper). -
    ISBN 0-226-30632-1 (paperback : alk. paper)
    1. Science and state—United States.
    2. Science—Political aspects—United States. I. Title.
Q127.U6G68   1999
509′.73′—dc21   98-52458
                                                    CIP

The Sciences having long seen their votaries labouring for the benefit of mankind without reward, put up their petition to Jupiter for a more equitable distribution of riches and honor. . . . A synod of the celestials was therefore convened, in which it was resolved, that Patronage should descend to the assistance of the Sciences.

SAMUEL JOHNSON, *Rambler,* No. 91, JANUARY 29, 1751

Contents

Foreword to the 1999 Edition

Sir John Maddox

Why should one now take time to read a book first published more than thirty years ago and which, in 1967, was not received as a work of such philosophical or historical importance that its meaning would survive the passage of the decades? After all, *The Politics of Pure Science* was originally a description of the mechanism for supporting basic science that the U.S. government had stumbled on by about the mid-1960s, together with an analysis of how well that mechanism then functioned. (Not *very* well, but in an interestingly deficient way: high-sounding declared objectives were routinely undermined by unspoken motives.)

So why, busy committee members will ask, should they read a book about the 1960s when they do not even have the time to read the agenda papers for tomorrow's meeting—there is always one—at which they will decide the fates of a dozen brave new ventures, the fruits of the imaginations of the brightest in the land? Similarly, the mechanism's customers (academic scientists and people like that) are likely to protest that what time they have to spare from the administration of their current research grants had better go to the still incomplete application for the next; if there is time left over, a few minutes in the laboratory would be a better investment than the reading of an historical text. Senators, congressmen, and parliamentarians generally have more winsome excuses. But taxpayers, a class that includes all the others cited, should pay close attention to what Dan Greenberg had to say all those years ago.

There are several reasons for doing so, one of which is historical. The first part of Greenberg's book, with its account of how a mechanism of public support for privately conducted research grew up to meet the needs of the postwar United States, goes to the root of the mechanism now in being, not only in the United States but wherever governments emulate the enormous achievements of U.S. research. What we forget until Greenberg reminds us is how the system was shaped by powerful individuals—Vannevar Bush, James Conant, and the like—buoyed up by the great technical achievements of World War II. There was no master plan, but almost a license for the erstwhile denizens of Los

Alamos and the Radiation Laboratory to carve out their own piece of the post-war action. That, for example, is the spirit in which Isidor Rabi set up Associ-ated Universities, Inc., to found and operate the Brookhaven National Labora-tory. Then, in this expansive climate, came the first Sputnik in 1957; optimism was suffused with panic that persisted until the Berlin Wall came down more than three decades later.

Those were the days when plurality flourished. Every agency in Washington seemed to want to fund a piece of the lustrous pie from which flowed recogni-tion (for the agency as well as its beneficiaries), influence (over other agencies) and even the occasional Nobel Prize. At about the time that *The Politics of Pure Science* appeared, I asked George Weil, then in charge of extramural programs at the Office of Naval Research, how he justified to the Navy the office's gener-ous support for research on the behavior of porpoises and in molecular biology. He put his tongue firmly in his cheek and asked, "Which of us can tell when we'll have porpoises flying U.S. Navy planes, or perhaps we'll go straight to RNA molecules?" (There was then a vogue for believing that the nucleotide se-quences of RNA might encode human memory.) At about the same time, I re-call a bizarre session of the Congressional Joint Committee on Atomic Energy at which the chairman, Representative Chet Holifield, tried to persuade Profes-sor Joe Weber (University of Maryland) that his attempts to detect gravitational waves, which had attracted huge newspaper headlines a few weeks earlier, had important implications for the nuclear industry and should be supported by the Atomic Energy Commission. (Weber's claims were premature and were soon shown to be unfounded.)

That points to a second reason why *The Politics of Pure Science* deserves another read. Within a few months of Greenberg's arrival at *Science* in 1960, he had given a new meaning to the concept of science policy. Previously, U.S. newspapers and magazines relied for their coverage of science on what emerged from the occasional reports of PSAC (the President's Science Advi-sory Committee)—one memorable document dealt with the plight of the Northeast's cranberry bogs—and from policy statements by the operating agencies, notably the AEC as it used to be, the Department of Defense, NASA (a post-Sputnik creation) and the stripling National Science Foundation, then only a bit-player. The National Academy's role in public affairs was largely confined to that of the Committee on Science and Public Policy (COSPUP), then feeling its way in the outside world. Nor was the President's science advi-sor a source of opinion and comment. In President Eisenhower's time, George Kistiakowsky divided his time between Harvard and the White House and was as laconic as a Trappist. Even on the day early in the summer of 1960 when the newspapers were full of the news that Gary Powers's U2 had been shot down over southern Russia, he did little but lean against a mantelpiece in the Old Executive Building and mutter that it was "a bad business." Left to themselves, of course, governments the world over make the best of events and, if that is not possible, say nothing.

Greenberg's big success was to recruit Congress to keep the agencies honest, or at least to get at the truth. He energetically exploited the way that congressional committees could wring information from public officials (and their underlings) in charge of operating agencies such as NASA and NIH: the process quickly revealed the tensions and discontents beneath the complacent facade the agencies sought to present in public. But congressional hearings are also notoriously occasions on which the agencies' customers, grant recipients for example, will tell Congress (and the world) of their discontents. Greenberg was not the first to recognize that Congress through its procedures had become a rich source of information on technical and scientific matters (Irwin Goodwin of *Physics Today* has worked this same source for decades); his achievement was to make vivid use of this torrent of information.

The illustration that sticks in the mind is Greenberg's invention of the character called Dr. Grant Swinger, a person meant to typify those whose evidence to congressional committees (called "testimony" on Capitol Hill) falls short of the standards of objectivity and dispassion of which the research professions are fond of boasting. Scientists are not, of course, the only witnesses at congressional hearings who magnify their own professional importance, who regard their own opinions as statements of the obvious or even as facts, and who seek to create the impression that the work on which they are personally engaged is not only of great national importance and of potentially huge benefit to humanity at large but is also, by good fortune and as a consequence of the witness's perspicacity, just at the point of fruition, at least if a sufficient supply of dollars is forthcoming as a result of the action a wise and farsighted congressional committee will undoubtedly now take. Dr. Swinger's discreet exaggerations were hilarious, deliberately so. The underlying message was deadly serious. The logrolling scientist-witness may not have disappeared altogether, but such a person would be more common (or recognizable) today were it not for the image of Dr. Swinger in people's minds.

On reflection, there should have been no great surprise that Congress proved to be such a powerful foil for Greenberg's science reporting. Does not the Constitution assure Congress equal weight with the Administration in the conduct of public affairs, and with the all-important right to approve the annual budget, line by line? But Greenberg's contributions to *Science* throughout the 1960s had an electrifying effect on rival publications (*Nature* included), which followed Greenberg as best they could. It is even possible that the closer attention paid by the interested press to the congressional committees that deal with technical and scientific matters may have enhanced their own opinions of the importance and seriousness of the inquiries they undertook.

The essential message of *The Politics of Pure Science* is its title: even academic research is everywhere inextricably bound up with the political process if for no other reason than that government subventions must come from government revenues and must be allocated by some mechanism or other. The U.S. system has the virtue of being more transparent than most others, with the possible exception of those in Nordic countries and the Netherlands. The research community has now come to terms with this once unpalatable truth, even if it has not yet learned to urge its claims on Congress in a manner that is at once effective and seemly.

How well has the system described in *The Politics of Pure Science* served the United States in the succeeding three decades? On the face of things, despite the flaws, the products of U.S. research laboratories during that period have been quite remarkable. To judge from the torrent of research publications in the major journals, the United States is the source of perhaps two-thirds of all the intellectual innovations that make contemporary science as exciting as it is. That may be proof that Vannevar Bush was right to describe *The Endless Frontier* (part of the title of his report to the President in 1945) as an inexhaustible source of benefit to mankind, but it does not prove that the system as it has evolved is as excellent as the results suggest.

There are several difficulties, some of them inherent to the U.S. Constitution and its consequences. The mechanism described by Greenberg in 1967 has three essential elements. Government support for research must ordinarily be by means of project grants to individuals, not to institutions, because the federal government has no direct role in education. Judgments among competing claimants for research funds must be made by peer review, administered by some branch of the federal government. And a diversity of funding sources is necessary to ensure that independent agencies can sponsor the research they believe they need. All three elements of the system are now being tested by external pressures more severely than in 1967.

The competitiveness of the project grant system is a large part of the explanation for the productivity of the U.S. research system in the past three decades. The fear that a grant will not be renewed keeps graduate students and postdoctoral fellows in their laboratories all hours of the day and night. But the price of failure can be that talented people are abruptly thrown into an uncertain labor market, which is cruel and which may also be a waste of talent. (There may just now be no difficulty in molecular genetics, but what happens to paleontologists, say?) The endemic sense of competitiveness almost certainly contributed to the rash of misconduct cases in the 1980s. It has also helped to erode the civility that once governed relations between competing laboratories and individuals. Unfortunately, the only way of blunting these pressures is to fund institutions as well as people, which is against the rules. (Germany, with a similar separation of powers between federal and state governments, nevertheless manages joint funding of research.)

Peer review is another source of difficulty. When rejection rates are high (as now), the process must fall short of Solomonic judicial standards. While there is no reason to believe that reviewers of research grant proposals are systematically prejudiced, they cannot fail to be influenced by a person's public reputation or that of his or her institution, which are to some degree legitimate components of the decision. More serious is the spurious weight given to publications, which in turn compels researchers to compete for space in a handful of well-known journals.

Pluralism is outwardly beneficial to the research community in that an application turned down by one agency may succeed when sent elsewhere, although the guidelines that determine the spending patterns of the different agencies now mesh together with less overlap than in 1967. More serious is the diversity of operating styles between the agencies: the slow adaptation of the Department of Agriculture to the climate of competitive grants is a striking testament of its devotion to its traditional dependents, the land grant colleges, for example. Yet the conflicting interests of the agencies have not stifled flexibility as might have been expected: who would have thought, in 1967, that Los Alamos would have an important role in the Human Genome Project? It remains a curious truth that all federal governments have repeatedly sought to create interagency coordinating committees for science and technology, and have been as often disappointed in the outcome.

All these difficulties were foreseen in 1967. Others, such as the way in which the United States has gotten a bad name in international collaborations for its inability to commit funds for more than a year ahead, are problems that have more recently come to light. So too are the two most serious problems on the horizon: the danger that the independence (or otherwise the usefulness) of the research universities will be eroded by the growing involvement of researchers in the management of commercial corporations and, linked with that, the emerging tension between academic and commercial researchers over questions such as access to or the patenting of genetic sequence data. The endless frontier is also an inexhaustible source of problems, few of them foreseen.

Greenberg's long view of the research enterprise in *The Politics of Pure Science* at least makes intelligible the uncertainties—and the excitement—of the past thirty years. And then there is another reason for spending time with this volume: it is a model of lucidity.

Foreword to the 1999 Edition
Steven Shapin

In the mid to late 1960s I was a graduate student in genetics. I was not a very dedicated student. I spent too much time in extraneous activities and not enough time preparing my culture medium and counting my fruit flies. I loved science (I still do), but, if the distinction is permitted, the day-to-day life of the experimental laboratory did not thrill me. Those extraneous activities on which (I was told) I was wasting my time included armchair theorizing about the nature of science—I had just read Thomas Kuhn's *The Structure of Scientific Revolutions* and found it exhilarating—and low-level participation in Scientists against the Vietnam War.

More mundanely, I was spending far too little time doing science and far too much time reading *Science,* the weekly magazine of the American Association for the Advancement of Science. Worse yet, the bits of *Science* I found most intellectually stimulating weren't the technical reports but the "News and Comment" section. I held journalists in high esteem—many Americans then did— and I thought that this was journalism at its best. There were a number of fine News and Comment writers at the time—I very much appreciated reporting by Elinor Langer, John Walsh, Luther Carter, Bryce Nelson, and Philip Boffey— but week-by-week I found myself most looking forward to pieces signed by D. S. Greenberg.

Dan Greenberg's writing was—and remains—wonderfully vigorous and acerbic. In 1966 he wrote that the National Science Foundation has "an instinct for trouble-avoidance derived from 15 years of congressional badgering, budget slashing, and allegations that NSF's clients have been living it up or pursuing esoteric nonsense at the taxpayers' expense," and he described the University of California at Berkeley as "the academic world's leading convalescent."[1] When Lyndon Johnson began seriously squeezing the federal budget for basic scientific research—partly as a result of the financial demands of the Vietnam War—Greenberg wrote that "scientists are among the more insecurity-ridden wards of the federal treasury," and that "a shrewd salesman might have prospered by offering mourning bands for lab coats."[2] And when the space program began casting about for ways to justify its massive costs by its modest civilian technological "spin-offs," Greenberg noted that "primitive cost-effectiveness analysis long ago discredited the burning of barns for the roasting of pigs."[3]

Greenberg's journalism could be very funny indeed. He had just launched the career of one of the great comic (and only semifictional) characters of the modern academy, Dr. Grant Swinger: the scientist on the make and on the plane to Washington; Prof. Morris Zapp in a lab coat; Director of the Breakthrough Institute and Chairman of the Board of the Center for the Absorption of Federal Funds (motto: "As Long As You're Up, Get Me a Grant"); recipient of the prestigious Ripov Award and the Segmentation Prize, annually given for the most publications from a single piece of research.

In 1966 Dr. Swinger delivered a speech at the Center for Intellectual Evanescence where he received the Hunter von Tenure award for his monograph "Overhead and Underhand: The Economics of Research." As Greenberg reported, Swinger applauded the President's decision "to teach an animal to speak in this decade," which was, as we all now know, the historical origin of NASA (the National Animal Speech Agency). Ah, "but which animal? And what should the animal be taught to say?" Nor, Swinger warned, should we be discouraged by those who suggest that no concrete benefits to human welfare are likely to come from such research: "Let us recall that it was Benjamin Franklin, or possibly Benjamin Spock, who said, 'What good is a baby?' And I believe it was Faraday who, when asked about nuclear fission, said, 'Someday you'll tax it.' I think the lessons of history are there for us to read. Let us hope we can read them clearly."[4]

As always with a really good joke, the point was deeply serious. And if the systematic seriousness of purpose was not evident in Greenberg's weekly reporting, it was inescapable in *The Politics of Pure Science* (*PPS*), which appeared late in 1967. There is a tension that lies at the very heart of the relations between scientific expertise and the democratic American state. It is a tension that results from the interplay between two sentiments that are, in our culture, equally persuasive and equally politically virtuous. On the one hand, we have the conviction that there are certain conditions in which scientific inquiry can best acquit its intellectual goals and thereby make its most effective contributions to human well-being and national interests. If the goose is going to lay the golden egg, it must be fed on demand with the richest possible cash diet and otherwise left alone. If you want reliable scientific knowledge, the price is a significant chunk of public funding and relief from the accountability to which other recipients of public funding are routinely subject. "Scientific research," the physicist Kenneth M. Watson wrote, "is the only pork barrel for which the pigs determine who gets the pork."[5]

On the other hand, democratic societies have legitimate reasons to worry about such states of affairs. The absence of accountability is a reliable sign of the absence of democracy. And however worthy the recipients of such special treatment may be, democratic sensibilities are thereby wounded to the core. When the experts are "on top," as opposed to "on tap," political decisions are redefined as technical decisions, and the cast of participants to decision-making is narrowed accordingly. We may have no practical choice but to defer to relevant scientific expertise when it tells us about the chemical composition of DDT, but lay members of democratic societies cannot be asked to defer mutely when it is proposed to spread DDT about in their neighborhoods. And we have no choice but deference to what physicists tell us about leptons and hadrons, but we are entitled to our say about whether several billions of our dollars are better spent right now on atom smashers or schools.

That is just the predicament in which the scientific community in a democratic society finds itself, and, as with many other dilemmas of complex modern life, the only sure mistake would be to think that some tidy resolution can be found. We defer to scientists in their proper domains because we have no alternatives, and because we are wholly convinced that theirs is the best natural knowledge going. Public belief in what scientists consensually have to say about their bits of the natural world is the most authentic act of homage: other forms of deference are either illegitimate or trivial by comparison. And we hold scientists to account as best we can when there are other worthy claimants on the national treasure, when what they propose to do conflicts with legitimate lay interests, and when we have reasonable grounds to suspect that they are not using the national treasure in the most responsible and efficient manner. Nor is there any moral or political obligation that the public and its political representatives meekly accept expert protestations that cuts, or even restricted growth, in funding for science necessarily jeopardize the national well-being or count as an affront to the Ideals of Western Civilization. The tension is just irresolvable: if we propose to resolve it in favor of minute political accountability, we are likely to stunt scientific development and the free play of creativity; if we exempt the scientific community from normal patterns of accountability, we compromise democracy through a tyranny of the experts, and ultimately create the conditions in which the scientific community comes to be resented and mistrusted.

Dan Greenberg's journalism, and his *PPS,* were not just powerful *reports* on this state of affairs, they were also powerful actions *within* it. "The politics of science," Greenberg wrote, "is in essence no different from other politics."[6] *PPS* showed that modern American science had become an authentic, and a considerable, element of modern American politics. Big Pure Science had a big budget, and, like other big-budget items, it had its vested interests, its lobbying apparatus, its pork, its public-relations exercises, and even what Greenberg frankly and repeatedly called its "ideology." To have a claim on the national purse, the "ideologists" of pure science had to do politics along the same lines as any other special-interest group, but they had the particular, and exceptionally interesting, task of publicly showing that a type of research often advertised as uniquely useless was in fact a legitimate claimant on the taxpayers' money.

The condition of proper accountability is that the public *know this,* know what the politics of science is like and how it came to be like that. The politics of science ought to be made as *visible* as other forms of politics, and making such matters visible was itself a virtuous political act. American Big Science was all grown up now, and, while the tradition of science journalism was to see its role as a cheering section, Greenberg reckoned that the proper place of science in a democratic polity could only be served if science was, and was seen to be, called to account in the same way as other recipients of public largesse. No one has ever performed that function as well as Dan Greenberg, and depressingly few science journalists have even tried.

PPS opens with a distinction: Greenberg says his task is "to write not about the *substance* but rather about the *politics* of science." Since the early 1970s I have earned my living as an academic historian and sociologist of science. In my particular line of work a lot has changed since *PPS* was written, and some of the historical and social scientific resources on which Greenberg then drew look very dated indeed. So far as my day job is concerned I belong to the sect of academics—often derisively called 'fashionable'—which maintains that the making, maintenance, and distribution of scientific *knowledge* is itself political. What we mean when we say this is that the work involved in making scientific knowledge is social (that is, collective), that there are no impersonal and global standards that regulate scientific judgment (that is, judgment always contains a conventional and local component), and that the credibility of scientific claims is never a matter of pure evidence or pure logic (that is, it always involves rhetoric and persuasion and is, in this sense, rightly called political).

A few of us in this line of academic work see our research as having—albeit in a very modest way—the same sort of inspiration as Greenberg's journalism. We want to make visible the processes by which the *substance* of scientific knowledge is made. Mostly we want to do this because we think this is the best way to meet normal standards of disinterested and searching academic inquiry. But some of us also believe that such visibility is good for a society that depends so much on scientific expertise and also good for the legitimate standing of science in a democratic society. Academic historians and sociologists who choose to write for a larger-than-usual audience can make such a contribution and so can the small number of science journalists who take as one of their tasks showing the public what scientific knowledge is like, not just as a finished product, but in the making.[7]

Of course, when sociologists of scientific knowledge clock off their day jobs, they know as well as anyone else what the distinction between substance and politics is all about: it's a way in which we might choose to say that we're interested just now in the funding arrangements for particle accelerators as opposed to the arrangement of particles within the nucleus. Such distinctions—like that between "science" and "society"—are resources around which important aspects of modern cultural life are organized, and in that sense *PPS* is not dated at all. It continues to be one of the most attractive models of how an important aspect of modern science might be made publicly visible and thereby publicly accountable.

Greenberg's journalism took as much courage as it did dedication and skill. His reporting won the respect of many leaders of the scientific establishment and the hostility of others. Some, indeed, condemned *PPS* as an insolent act of *lèse majesté*. The physicist-reviewer chosen by *Science* itself said that its "overall effect is to demean";[8] the *Scientific American* reviewer deplored what he saw as Greenberg's tendency to "emphasize the negative": this was said to be dangerously irresponsible journalism when the scientific community was in such a "precarious position."[9] *The Bulletin of the Atomic Scientists* reviewer—an NSF Program Director—slapped Greenberg's wrists for being "impolite," noting that "most scientists will be unhappy indeed . . . with its apparent working premise that scientists are no more noble of purpose and performance than are ordinary men."[10] Greenberg survived unscathed, but it was a warning to other journalists who might think that holding the scientific community to account was risk-free.

There was also something about the precise historical juncture at which Greenberg wrote *PPS* that made his undertaking both possible and pertinent. In the mid to late 1960s American science was both tremendously secure and increasingly nervous about its standing in the culture and *vis-à-vis* the State. The legacy of penicillin, hybrid corn, and especially the Manhattan Project was a generation wholly satisfied that pure science was both a major adjunct to national security and a contributor to the national welfare. Utilitarian justifications for even the most apparently disengaged forms of basic research fell on fertile ground, while the Cold War ensured that the same sensibilities that identified nightmarish missile gaps might also benefit American physicists wanting bigger accelerators than those possessed by the Soviet Union.

Scientific research might, in principle and under such conditions, expand to absorb all the resources the nation could provide. It was not Lyndon Johnson but Nikita Khrushchev who said that scientists "always want more than they can get—they are never satisfied." "Dissatisfaction," Greenberg wrote, "is the universal base line in science money matters."[11] And it was in that setting that Derek deSolla Price famously drew the exponential growth curve predicting that, were present patterns of growth to be maintained, a time would soon come when there would be "two scientists for every man, woman, child, and dog in the population, and we should spend twice as much money as we had."[12] That was never going to happen for reasons of principle, and it was certainly not going to happen in the late 1960s when the Vietnam War and the Great Society were conducting a tug-of-war for limited federal funds, a contest in which the crunch truly arrived for the American scientific establishment. Greenberg probably never saw his role as beating the drum for research dollars, and, equally probably, he never doubted the legitimacy of scientists' claims on a healthy share of those dollars. But he could not make himself a party to those who predicted national disaster if research funding was somewhat less than the scientific community could absorb: "Relative to other segments of American society—let's face it—research is not, nor is it ever likely to be, badly off. Thus, if one is shedding tears for those who must tithe for the costs of the Vietnam War, the post-doc who lost a trip to a foreign conference might properly take a place behind the slum kid who had a Head Start program budgeted out from under him."[13]

The American culture of the 1960s seems to have accepted the ultimate cultural, social, and economic value of science as a matter of course. At the same time, the culture also began to accept that the scientific enterprise could and should be held up to disinterested and responsible scrutiny. The Vietnam War raised consciousness about the moral responsibility of scientists and the closeness of the relationships that had built up between the universities and the military. On the other side of the political spectrum, the Pentagon's Project Hindsight (1963–66) marked emerging official skepticism about the tight causal links between basic research and technological innovation that had been central to public justifications of science.[14]

Myths about the nature, conduct, and social relations of science were then being challenged by both academic and popular commentary: Thomas Kuhn's

The Structure of Scientific Revolution (1962) took on simplistic rationalist legend about scientific judgment and change, subjecting it to the harsh scrutiny of warts-and-all history; James D. Watson's autobiographical *The Double Helix* (1968) gave voice from within the scientific community to growing cynicism about high-minded conceptions of the scientist's vocation; Jacques Barzun's wonderful, and now undeservedly neglected, *Science: The Glorious Entertainment* (1964) claimed that "Science and scientists would benefit from not having to sustain a myth of perfection"; and Rachel Carson's *The Silent Spring* (1962) shook public confidence in the wisdom of blindly accepting expert judgments of environmental risk. None of these writers considered that science could be well served by perpetuating Golden Age myths or by exempting science from all aspects of criticism. The scientific spirit of disinterested curiosity and forthright criticism could and should legitimately be applied to science itself. Academic and journalistic writing responded in their different ways to a historical moment in which you could be critical of many aspects of science, and associated science-myth, without the cry of "anti-science" automatically going up.

The Politics of Pure Science is, therefore, a historical document, but it is a document of unique importance: its sensibilities are enduring; it recalls a time of special significance for the understandings we now have of the relations between science and society; and it reminds us how consequential science journalism of this kind and quality can be.

NOTES

1. *Science,* October 7, 1966, p. 130.
2. *Science,* November 4, 1966, p. 618.
3. *Science,* September 1, 1967, p. 1018.
4. *Science,* December 16, 1966, pp. 1424–25. Grant Swinger made his original appearance in *Science* in 1964, and intermittently graced the pages of that journal, as well as the *New England Journal of Medicine,* the *Washington Post,* and, after Greenberg left *Science* in 1971, his proprietary *Science and Government Report.* A selection of *The Grant Swinger Papers* was privately published in 1981 and an expanded second edition in 1990. If you miss Dr. Swinger, or, especially, if you are too young to have even heard of him, join me in urging Greenberg and the University of Chicago Press to bring him to life once more.
5. Quoted in *PPS,* p. 151.
6. *PPS,* p. xxv.
7. For a token of such academic commitment, see H. M. Collins and Trevor J. Pinch, *The Golem: What Everyone Should Know about Science* (Cambridge, U.K.: Cambridge University Press, 1993). Occasionally, science journalism for *The Economist* has attempted to show

readers the uncertainty and contingency of much scientific judgment at the research frontier; see, for example, a fine series on the conduct of scientific experiment titled *Tests of the Truth: The Experiment in Modern Science,* Oliver Morton and Geoffrey Carr, eds., and published in *The Economist* between November 1992 and February 1993.

8. Frank T. McClure, "Outside the Laboratory," *Science,* May 17, 1968, pp. 752–53.
9. Victor F. Weisskopf, "Is Pure Science Doomed to Decline in the U.S.?" *Scientific American* 218, no. 3 (March 1968), pp. 139–44.
10. Joel A. Snow, *Bulletin of the Atomic Scientists* 24, no. 5 (May 1968), pp. 34–36.
11. *Science,* October 13, 1967, p. 230; see also *PPS,* p. 156, second footnote.
12. Derek J. deSolla Price, *Little Science, Big Science* (New York: Columbia University Press, 1963; reprinted 1968), p. 19.
13. *Science,* October 13, 1967, p. 230.
14. *Science,* November 18, 1966, pp. 872–73; also Raymond S. Isenson, "Project Hindsight: An Empirical Study of the Sources of Ideas Utilized in Operational Weapon Systems," in *Factors in the Transfer of Technology,* William H. Gruber and Donald G. Marquis, eds. (Cambridge, Mass.: MIT Press, 1969), pp. 155–76.

Acknowledgments and
a Note on Sources

The recollections and opinions of many persons in and around the scientific community went into the preparation of this book. Since my quest for materials ranged from tape-recorded interviews of several hours to brief conversations, it would be difficult to acknowledge all of the hundreds of persons who generously assisted me, but in particular I would like to express my appreciation for the help of the following:

Luis Alvarez, Allen Astin, Willard Bascom, the late Lloyd Berkner, Detlev Bronk, Harvey Brooks, Vannevar Bush, William D. Carey, Harrison Brown, Emilio Q. Daddario, Lee DuBridge, Sidney Farber, the late John Fogarty, Delphis Goldberg, Fred Harris, Caryl Haskins, Leland Haworth, Harry Hess, Donald Hornig, James R. Killian, George B. Kistiakowsky, Arthur Kornberg, Joshua Lederberg, W. F. Libby, Edwin M. McMillan, the late Clark Millikan, Walter Munk, Joseph Murtaugh, W. K. H. Panofsky, Emanuel Piore, Edward Purcell, I. I. Rabi, Norman Ramsey, Isadore Ravdin, David Z. Robinson, John Rubel, Matthew Sands, Frederick Seitz, James Shannon, Harold Urey, Bernard Waldman, Alan T. Waterman, Warren Weaver, Jerome B. Wiesner, John T. Wilson, and Jerrold Zacharias.

With two general exceptions, I have indicated the sources of the material I have employed. First, some materials were furnished on a confidential basis, with the understanding that they might be used but the source might not be revealed. Where I judged that the value of the material outweighed the liability of concealing the sources, the decision went to using the materials in accord with the conditions that accompanied them. Second, since numerous quotations in the text are derived from interviews with many of the persons listed above, I have dispensed with source citations in such cases. So, let it be understood that when contemporary figures are quoted without citations, the quotations come from these interviews—in most cases, tape-recorded interviews.

I also wish to express my appreciation to a number of persons who have reviewed all or parts of the manuscript and who have made many helpful suggestions. These are Philip H. Abelson, Norman Kaplan, Elinor Langer, Harold Orlans, Don K. Price, David Z. Robinson, Seth Tillman, and Dael Wolfle. The assistance of Mrs. B. Drummond Ayres, Jr., and Mrs. Douglas Parrillo was indispensable in the preparation of the manuscript. I also wish to thank Professor Harry Woolf, chairman of the Department of History of Science at Johns Hopkins University, and his wife, Pat, for their friendship and encouragement while I was working at Hopkins. The Carnegie Corporation has my gratitude for the financial assistance it provided me while I was on part-time leave from *Science* to work on the book. As readers of this book already know, or will find, much controversy swirls around the grant business, but, in the best and purest tradition, Carnegie paid my way and left me alone.

Finally, I would like to express my deepest gratitude to Dael Wolfle, the executive officer of the AAAS and publisher of *Science,* and Philip H. Abelson, editor of *Science.* Under their leadership, the *Science* news department has become a unique vantage point for scrutinizing the politics of science. Without the view afforded by that vantage point, this book would have been far more difficult, perhaps impossible, to complete.

I, of course, add the usual disclaimer: I, and none of the persons or institutions cited, bear responsibility for whatever views, errors, or deficiencies follow.

Introduction

In 1961, I joined the news department of *Science,* the weekly magazine of the American Association for the Advancement of Science. My assignment was to write about scientific affairs, but it was a particular aspect of science that I was to deal with. I was to write not about the *substance* but rather about the *politics* of science.

There is, of course, a great difference between the two, though they are not unconnected. The substance of science is in the procedures and results of research and experimentation. The politics is in the people, institutions, and processes that determine what is to be researched, by whom, where, under what circumstances, and, finally, what is to be done with the results. The substance is centered in the laboratory. The politics is centered in the committee room.

The substance of science is unique to science. But the politics—except to the extent that it is flavored by the peculiar traditions of science—is not. Science, like agriculture, the military, labor, business, or the civil rights movement, has its vested interests, elites, downtrodden, alliances, bosses, loves, and hates. The politics of science is in essence no different from other politics. It is a bit more obscure and, it seems to me, usually better mannered, though perhaps it would be appropriate to borrow a line from Wordsworth and describe it as "Nor harsh, nor grating, though of ample power to chasten and subdue."

In any case, this book is about the politics of science. But even this has to be refined further. It is about a particular slice of science—pure science, or, as it is often referred to, basic or fundamental research. There are two reasons why it is confined to pure science. The first is that my years with *Science* have been concentrated on the politics of pure science, and this book draws heavily upon the knowledge that I accumulated during those years. The second is that the politics of "non-pure" science—that is, applied science and technology—has got a good deal of very capable attention in the press and in various book-length studies. Space, atomic energy, military technology, and so forth do not lack attention. On the other hand, though I have encountered many footprints during my researches, I believe there are some substantial areas in the politics of pure science that heretofore have not received the attention they merit, and that is why I presume to add to the already overswollen inventory of books on science and government.

Reading through substantial portions of this inventory leaves one with the impression that two schools of thought predominate on the subject of science and government. The first I would call "the immaculate conceptionist," and the second "the corruptionist." The former generally sees wisdom, goodness, and inevitability in the relationship that has flourished between science and government since World War II. The latter sees a Treasury raid from within and a usurpation of the democratic process.

To sample the immaculate thesis, we can look to the words of Harvey Brooks, dean of engineering and applied physics at Harvard, who writes that "among basic scientists there is a stern ethical code associated with the question of scientific credit and priority which is powerfully sanctioned within the scientific community. It provides a model for a similarly sanctioned, though unwritten, code with respect to the objectivity of scientific advice for government. . . . If it can be said that any policy viewpoints on science-related matters have become dominant in government, this has been imposed more by the logic of events than by any particular group of advisers." *

On the other side, there is Robert M. Hutchins, former president of the University of Chicago, who states: "My view, based on long and painful observation, is that professors are somewhat worse than other people, and that scientists are somewhat worse than other professors. . . . There have been very few scientific frauds. This is because a scientist would be a fool to commit a scientific fraud when he can commit frauds every day on his wife, his associates, the president of the university, and the grocer. . . . A scientist has a limited education. He labors on the topic of his dissertation, wins the Nobel Prize by the time

* Harvey Brooks, "The Scientific Adviser," *Scientists and National Policy Making,* Robert Gilpin and Christopher Wright, eds. (New York: Columbia University Press, 1964), p. 95.

he is 35, and suddenly has nothing to do. . . . He has no alternative but to spend the rest of his life making a nuisance of himself." †

My intent is not to adopt or pursue either of the themes stated above, but rather, without reverence or cynicism, to look into the politics of pure science to determine what it is and how it works.

D.S.G., WASHINGTON, JULY 1967

† Robert M. Hutchins, *Science, Scientists, and Politics,* an occasional paper on the role of science and technology in the free society, Center for the Study of Democratic Institutions, 1963.

BOOK
ONE

I

The Scientific Community

There isn't a scientific community. It is a culture. It is a very undisciplined organization.

I. I. RABI, Nobel Laureate in Physics, 1965

Science, in its pure form, is not concerned with where discoveries may lead; its disciples are interested only in discovering the truth.

ALAN T. WATERMAN, Director of the National Science Foundation, *American Behavioral Scientist*, December 1962

In studying the American scientific community, it would be useful to discover a scientific establishment, a group small in number, associated with venerable institutions, that commands affairs regardless of what the *pro forma* processes may be. A cursory examination of the community suggests the existence of such an establishment, but to the extent that it exists, I think that emphasis upon it can mislead more than enlighten in understanding the politics of science. Therefore let us begin with a paradox: There is no American Scientific Establishment. Yet Harvard, MIT, Caltech, and the University of California are its Oxbridge. Two World War II research centers, the MIT Radiation Laboratory and the Los Alamos Scientific Laboratory, of radar and atom bomb fame, respectively, are its Eton. The Cosmos Club in Washington is its Athenaeum, the physicists are its aristocracy. The National Academy of Sciences is its established church, and the President's Science Advisory Committee is its Privy Council.

The Establishment has two branches. The eastern wing is headed by a triumvirate: Jerome B. Wiesner, science adviser to President Kennedy and Provost of MIT; George B. Kistiakowsky, science adviser to President Eisenhower, professor of chemistry at Harvard, and vice president of the Academy; and Detlev Bronk, president of Rockefeller University and former president of the Academy. The western wing is headed

3

by Lee DuBridge, another Eisenhower science adviser, who is president of the California Institute of Technology.

I. I. Rabi, still another Eisenhower confidant and emeritus professor of physics at Columbia, is the *éminence grise,* and *Science,* the weekly magazine of the American Association for the Advancement of Science, is the quasi-official journal. The National Science Foundation is the bank; Frederick Seitz, president of the Academy, is the peacekeeper; and Emanuel R. Piore, vice president of IBM, is the chief trouble-shooter. Harvey Brooks, dean of engineering and applied physics at Harvard, is the chief political theoretician. J. Robert Oppenheimer is the tragic hero.

Illusion or reality, I do not think the establishment concept provides much help for examining the politics of science. So, at the outset, I would like to dispel any hopes for elitist theories, compact tables of organization, or directions to a central command post. We are examining a very untidy subject.

The part of our society that is conveniently referred to as "the scientific community" is an agglomeration of people and institutions generally centered about workshops called laboratories and engaged in activities called research. This definition takes us only to the gates of the community. Once inside, it is common for the visitor to direct his attention to the substance of science, to what scientists are doing, to what their work is about. But we are concerned with something else— the political behavior of science, both in its relations with the nonscientific world and in its internal aspects. What we want to know is how it works, in terms of the people and the forces, traditions, policies and practices that determine who gets what. For this purpose, it is necessary, first, to look upon the scientific community not simply as the location for a particular type of activity; rather, it is necessary to look upon it as a way of life, as a collection of particular and persistent sensitivities, values, and vulnerabilities, and as a style of behavior distinct unto itself. And then, after having looked at these internal characteristics, we will be in a better position to examine and understand the community's dealings with other segments of our society.

If we are looking at the substance and practice of science, we find the common denominator is a dedication to the understanding of the universe through systematic investigation and measurement, through the harnessing of curiosity, training, discipline, and instruments. (Some would add intuition and good luck.) But if we direct our attention to the politics of science, internal and external, we find that the common denominator involves different matters.

From these two perspectives, the scientific community can be seen as bound together by a twofold ideology: first, a desire for society to support, but not govern, science; and second, for the community of science to exist as a loosely organized entity—meritocratic anarchy may best describe it—in which hierarchy and tables of organization bear little relation to the realities of power. This dual ideology is important, not because it governs events, but rather because it affects the vision of scientists and influences the stances that they adopt, both in relation to each other and in their dealings with nonscientists. In its starkest theoretical formulation, the ideology is contained in the assertion of Michael Polanyi: "The pursuit of science can be organized . . . in no other manner than by granting complete independence to all mature scientists. They will then distribute themselves over the whole field of possible discoveries, each applying his own special ability to the task that appears most profitable to him. The function of public authorities is not to plan research, but only to provide opportunities for its pursuit. All they have to do is to provide facilities for every good scientist to follow his own interest in science." [1] Derivations of this puristic concept are to be found in the statements of men very much concerned with the practical problems of administering the affairs of the scientific community. In the words of Alan T. Waterman, former director of the National Science Foundation: "Basic research is a highly specialized activity; it is not one where the judgment of laymen has validity. Consequently, planning for basic research and such evaluation of its performance as is needed for the continuation of existing programs must be left in the hands of competent and experienced scientists." [2] Or, consider the words of a group of chemists: "To understand how research is organized at the universities, one must realize that a university does not operate through a chain of command. Each member of the faculty is expected to develop a research program of his own. His research is not subject to review or criticism by anyone in his own department, except on his own initiative." [3] Polanyi, Waterman, and the chemists were addressing themselves more to what is desirable than to what is actual in the world of science, but in science, as in politics and religion, theoretical formulations can tinge perceptions, stir up passions, and move men to action.

If education and place of employment are the criteria, it is relatively simple to locate the American scientific community and identify the institutions through which it works. Statistics on the size and cost of science are harder to use, however, because it is difficult to pluck science from the mass of scientific and technical endeavors that are often lumped together as "science." First, a quick survey of the dimensions of the scientific community. In 1960, the latest year for which comprehensive figures are available, 335,300 persons gave "scientist" as their occupation, 822,000 said they were engineers, and 775,100 said they were technicians. Of the total who considered themselves scientists, 176,500 were employed by industry, 98,100 by colleges and universities, and 60,700 by federal, state, and local governments.[4]

These figures are grossly swollen, probably by self-flattering definition. If we are to locate the segment of the mass which conducts the politics of science, it is necessary first to seek out a particular identifying characteristic: possession of the doctoral degree, for the leaders almost invariably possess the advanced scientific training that precedes the degree. Again, definite figures are lacking. But a good assessment is to be found in the federal government's National Register of Scientific and Technical Personnel, which is based on a biennial survey that is believed to cover some 90 percent of all science doctorates, and 75 percent of all persons, with or without degrees, whose work calls for a scientific background. In 1964, according to the Register, the statistical profile of American science showed only 56,457 Ph.D.'s in the natural and physical sciences. (See Table.) The number of Ph.D.'s in engineering totaled only 10,000.[5] Now two important points. First, in terms of place of employment, educational institutions accounted for the largest group of doctorates in science, 47 percent; industry employed 29 percent, and government, 14 percent. Second, the gross number of Ph.D.'s is rather small, and when we deduct those at the end of their careers, plus those who hold degrees but do not practice science, we have to conclude that, while we hear a great deal about the incredibly rapid expansion of scientific research, real scientific research—as distinguished from technological activity—actually absorbs a rather minuscule portion of the nation's working force.

The rank and file of science, as is the case with the rank and file of any group or profession, goes about its business and generally leaves the politics to a handful of leaders. How big is this handful in the case of science? Christopher Wright, of Columbia, a longtime observer of the science and government relationship, estimates the number at 200 to 1000. James R. Killian, Jr., of MIT, who has been at the inner core

NUMBER OF SCIENTISTS, BY FIELD AND HIGHEST DEGREE, 1964

SCIENTIFIC AND TECHNICAL FIELD	TOTAL	Less than a bachelor's degree	HIGHEST DEGREE				No report of degree
			Bachelor's	Master's	Professional medical	Ph. D.	
			NUMBER				
All fields	223,854	2,878	72,364	61,222	5,925	79,372	2,093
Chemistry	63,053	590	27,377	12,229	368	21,789	700
Earth sciences	17,907	254	8,101	5,829	1	3,578	144
Meteorology	5,510	1,147	2,524	1,137	—	479	223
Physics	26,698	185	7,673	8,352	30	10,286	172
Mathematics	17,411	210	4,917	7,464	5	4,603	212
Agricultural sciences	9,526	63	4,362	2,676	11	2,367	47
Biological sciences	27,135	71	3,172	5,028	5,408	13,355	101
Psychology	16,804	4	417	5,464	52	10,843	24
Statistics	2,843	43	810	1,133	3	804	50
Economics	12,143	90	2,613	4,204	2	5,091	143
Sociology	2,703	3	65	434	9	2,179	13
Linguistics	1,351	1	162	407	2	729	50
Other fields	20,770	217	10,171	6,865	34	3,269	214

(Source: National Science Foundation, *Reviews of Data on Science Resources*, VOL. 1, NO. 2 [December 1964].)

of the relationship, uses the lower figure to describe the "consistently influential." [6] The numbers shift with time and issues, but there is no doubt that the politics of science is generally conducted by a remarkably small number of people.

Now, to move to other matters: the financial resources available to the practitioners of science and the institutional array in which these resources are employed. The amounts that this country now spends under the heading of "research and development" ("R and D"), and the rate of growth of these expenditures during the past twenty-five years, are spectacular. In 1940, the nation's total spending in these categories is estimated to have been $345 million.[7] The commonly used figure for the federal portion of that total is $74 million.[8] (Past and current figures must be regarded with a good deal of caution. Research defies precise accountability, and it is only recently that statistical studies in this area, principally under the auspices of the National Science Foundation, have been receiving significant support. The development of what is referred to as "research on research" is itself in large part an outgrowth of concern and dispute over the phenomenally rapid growth of federal expenditures for research. But it is a young and uncertain field of scholarship.) The 1940 total for research and development was approximately .03 percent of the gross national product, and the federal contribution constituted about .08 percent of the federal budget.[9] By 1962-63, the nation's total research and development expenditures had risen to $16.4 billion, of which the federal government provided $12.2 billion. The 1962-63 total was close to 3 percent of the gross national product, and the federal contribution constituted approximately 14 percent of the federal budget.[10] But again, as was the case with gross figures on employment in science, it is necessary to move on to more refined numbers to arrive at an understanding of the financial component of the politics of pure science.

In assessing the finances of what comes under the heading of research and development, it is difficult but necessary to attempt to locate boundaries between various segments of scientific and technical endeavor. And here we find that we have to be content with shadings rather than with precise demarcations. Let us start with an idealized compartmentalization, consisting of "development," "applied research," and "basic research." *

* The standard federal definitions of these three types of research are those devised by the National Science Foundation: "Basic research is research that

If development is the translation of existing knowledge into hardware, gadgets, techniques, or new material, then development accounts for the vast majority of R and D expenditures—somewhere around 68 percent.[11] An example of development would be the employment of existing knowledge about electromagnetic propagation to devise a new radio communications device. In general, this is the function of engineers.

If applied research is the quest for a new understanding that is specifically needed to make possible a new development, then applied research accounts for about 22 percent of R and D expenditures. Again taking radio communications as a case in point, an example of applied research would be the creation of a smaller, or perhaps less resistant transistor to meet the weight or power requirements of the radio device. This generally involves more science than engineering, but the two often tend to merge in this area.

If basic research is the quest for fundamental knowledge, regardless of the purpose to which it might be applied, then basic research is the smallest portion of R and D, amounting to only about 10 percent of the total. In the field of electromagnetic studies, its findings might ultimately be incorporated into communications devices.

But—and this is what a lot of the politics of pure science is all about —the basic researcher is not primarily, and perhaps not at all, concerned with utility. His objective is an understanding of fundamental phenomena, regardless of their utility. And there lies the fundamental political dilemma of basic research in the United States. Patronage, public and private, comes to basic research for many reasons—but the strongest reason is a belief that utilizable results may ensue. However, the motives of the recipients of basic research funds only partially overlap with the utilitarian motives of the patrons. The motivations, of

---

is directed toward increase of knowledge in science. It is research in which the primary aim of the investigator is a fuller knowledge or understanding of the subject under study, rather than a practical application, as is the case with applied research." "Development is the systematic use of scientific knowledge directed toward the production of useful materials, devices, systems, or methods, including design and development of prototypes and processes. It excludes quality control, routine product testing, and production." (*Federal Funds for Research, Development, and Other Scientific Activities*, NSF 65-19, pp. 56–57.) NSF, which, as we shall see, has had severe budgetary difficulties in fulfilling its mandate to support basic research and scientific training, takes care to point out in the foreword to this report, "Only basic research and certain broadly ranging components of applied research should be thought of as investments in science in the more general sense. For this work, costs are relatively small."

course, vary among individuals, but, in general, basic researchers engage in basic research because they find it a stimulating and agreeable way of life, not because it may ultimately lead to better television sets. This dichotomy in motivations was frankly acknowledged by Leland J. Haworth, director of the National Science Foundation, when he stated, "We [the scientists] . . . know the great cultural and intellectual value of science. But we are not good salesmen. The cultural argument, of course, competes with similar arguments for other fields of learning. And we would, in my opinion, be hard put to prove uniqueness for science in this sense. Large federal sums for culture's sake can come only when all culture is heavily supported. So for the present our best drawing card for financial support by the public is the ultimate usefulness of science. I do not defend the fact that this is so; I simply state it as a fact." [12]

In any case, pure science regularly strives to define itself and to distinguish its essential qualities from those of other technical activities. In my estimation, the task was accomplished long ago by two unrecognized students of astronomy, Huck Finn and his companion Jim, the runaway slave. Huck relates that when they were floating on their raft, "We had the sky up there, all speckled with stars, and we used to lay on our back and look up at them, and discuss about whether they was made or only just happened. Jim he allowed they was made, but I allowed they happened; I figured it would have took too long to *make* so many. Jim said the moon could 'a' *laid* them; well, that looked kind of reasonable, so I didn't say nothing against it, because I've seen a frog lay most as many, so of course it could be done. We used to watch the stars that fell, too, and see them streak down. Jim allowed they got spoiled and was hove out of the nest." [13]

Pure science is the institutionalization of Huck's and Jim's curiosity.

The next distinction in the national research panorama involves the distribution of money for research and development. And here we find a firm and durable pattern: the government provides most of the money, through contracts or grants, and industrial and educational organizations perform most of the work. Proceeding one step further, it is industry that performs most of the developmental and applied research, while universities are the largest single performers of basic research. Looking at 1961-62, when the national total for R and D was $14.7 billion, we find that the federal government provided 65 percent of the overall amount, industry 32 percent, and colleges and universities 3 percent. But when it came to using the money, 74 percent of the nation's research and development was performed by industry, 14 percent by government agencies, 6 percent by colleges and universities, and the remainder by private research institutes or administrative hybrids known as federal contract research centers—federally owned and financed research facilities operated under contract by nonfederal organizations, usually universities.[14] Thus, industry uses the bulk of the nation's expenditures for R and D, but when we look at the 10 percent or so of R and D expenditures that goes into basic research, we find that the pattern is quite different. About $1.5 billion was spent for basic research in 1961-62. The *sources* of this amount were: government, 57 percent; industry, 24 percent; colleges and universities, 12 percent; and other organizations, 7 percent. The *users* of the money were: colleges and universities, 39 percent on their main campuses, plus another 8 percent for operation of federal contract centers; industry, 27 percent; federal agencies, 16 percent, and other private organizations, 10 percent.[15] Such is a rough financial profile of contemporary American research and development. As we look in particular at the small but influential basic research segment, it is necessary, however, to make additional refinements.

First we must distinguish between "big science" and "little science." Big science is expensive science, involving large teams of scientists and technicians working with great facilities such as oceanographic vessels, nuclear accelerators, and radio telescopes. These cost a great deal to build and they cost a great deal to run. For example, the most recently completed piece of big science equipment, the two-mile-long "atom smasher" known as the Stanford Linear Accelerator, cost $114 million to build and consumes some $25 million a year in operating costs, all of it financed by the federal government. Little science, on the other hand, involves fewer people per project and far less costly equipment, the extreme in this category being the lone theoretical mathematician whose tools are paper and pencil. In substance science, big or little, is science, and there is as yet no rational method for correlating the cost of research and its intellectual worth or ultimate utility. But the economics of big and little science are strikingly different, and inevitably so are the politics.*

* In 1964, the federal government reports, it spent $273 million on basic research in physics and $93 million on basic research in chemistry. (National Science Foundation, NSF 65-19, p. 6.) No statesman of science would contend

Just as it would be convenient to discover a scientific establishment, it would be convenient to discover a rational pattern in the layout of the institutions of science, but no L'Enfant figured in putting together the scientific enterprise. The scientific leadership, furthermore, like inhabitants of a trackless jungle, find certain advantages in an amorphous landscape. Institutionally, the closest conformity to the substance of science is to be found in the scientists' professional societies. Since the establishment of the Accademia dei Lincei in Italy in 1603, such associations have existed for the purpose of communication, companionship, common defense, and furtherance of professional interest. By generous definition, there were in the United States in 1961 nearly sixteen hundred scientific and technical societies, ranging from the tiny and whimsical American Miscellaneous Society (which, as we shall see, was to become the center of some wholly unwhimsical matters) to the American Chemical Society and the American Association for the Advancement of Science, each with more than 100,000 members.[16] The societies are generally catchall organizations open to anyone with at least minimal qualifications in the profession. As such they are too diffuse to wield significant influence in the politics of science.

At the honorary apex of American science sits the National Academy of Sciences, of which we shall hear a great deal more. This is a tradition-bound, self-perpetuating body, with a current membership of approximately 780. Established during the Civil War through the machinations of a number of scientific luminaries who were disturbed by government's indifference to science, the Academy holds a congressional charter designating it as scientific adviser to the federal government. The charter, however, specifies that the Academy is to advise only when advice is sought, and throughout most of its century-long history, until quite recently, in fact, the government did not often ask.

To the numerical limits of its membership, the Academy is supposed

---

that the support of physics at three times the rate of chemistry is the deliberate consequence of a rationalized policy-making process. Rather, in seeking to explain the origins of the status quo, there is a tendency to allude to mechanisms remarkably reminiscent of the "invisible hand" of early capitalist economics. For example, Jerome B. Wiesner, science adviser to President Kennedy, put it this way in explaining science policy to a congressional committee: "Without really having planned it, in fact I would say without really understanding it, we have evolved a system, as we often do in this country, of checks and balances and interactions that gives us the strongest scientific and technological community you can find in the world." (*Government and Science, Hearings,* Subcommittee of the House Committee on Science and Astronautics, 1963, p. 74.)

to encompass the most professionally distinguished members of the American scientific community, but with the average age of admission being 49.1 and the average age of the total members 61.6,[17] there are many distinguished young scientists outside the Academy, and regardless of age, there are some rather undistinguished ones inside. On the whole, however, the Academy represents the best of post-middle-age science in America. Election is by a process that rivals the papacy for mystery. Through an infinitely complex screening process, those who are inside select 45 new members for admission each year, and, in general, they select with great care, for in the hierarchy of scientific honors, admission to the Academy is second only to winning the Nobel Prize. Nevertheless, despite the time, care, and occasional politicking that goes into the selecting Academy members, it is illuminating to note that since 1950, the Nobel Prize has been awarded to nine scientists who, at the time of the award, had not been admitted to the Academy. In all but two cases, Academy membership was then swiftly bestowed upon them. (One died after receiving the prize, and one still has not made it, though he got the prize in 1954.) [18]

Around the half-century mark in its history, which came during World War I, the Academy recognized that its honorary and advisory functions did not mix well, and, accordingly, it established the National Research Council to draw upon the scientific community at large for advisory services. Today NRC annually obtains the services—without cost, except for travel expenses—of some 4000 scientists and engineers throughout the country, to provide advice on an incredibly broad array of matters, ranging from the nutritional requirements of laboratory animals to selection of a site for the most costly research center ever constructed.

The proper name of the overall organization is the National Academy of Sciences–National Research Council (NAS-NRC). Appended to this, in a loose fashion, is the recently established National Academy of Engineering, the creation of which, several years ago, underlined the fact that NAS-NRC had become almost wholly a creature of pure science. The engineers, writhing from inadequate representation, were initially going to set up their own independent academy, but eventually agreed to begin their existence in confederation with the scientist-dominated Academy. So far, the arrangement has worked out fairly well.

The power and influence of the Academy as an institution are open to question. It has been called the Supreme Court of Science and science's League of Women Voters. But, whatever the role of the institution may be, there is no doubt that the internal politics of science is heavily weighted with Academy members. Academy membership is not a prerequisite for entry to the political arena, but membership certifies scientific accomplishment, and scientists generally prefer to entrust their political affairs to those who have demonstrated scientific or technical creativity. Occasionally, a career administrator, such as Killian of MIT, will go to the top in the politics of science, but, on the whole, the ranking politicians, such as Wiesner, Seitz, Rabi, DuBridge, Kistiakowsky, and Bronk, first made their mark in the substantive work of their professions, and then moved into the politics. These moves, it is important to note, often carry them into a series of interlocking positions. Thus, Seitz not only is president of the Academy, but also chairs the Defense Science Board, which is the highest ranking science advisory body in the Pentagon, and he is also a member of the President's Science Advisory Committee. When Bronk was president of the Academy, he sat on PSAC and also chaired the National Science Board, which is the top advisory board of the National Science Foundation.

Decorum, solemnity, and tradition pervade the Academy's marbled, banklike headquarters building on Washington's Constitution Avenue. But, just as a peaceful seascape can conceal violent struggle beneath the waves, the Academy's placid facade often serves to cover mighty struggles within the scientific community. Thus, in 1950, when James B. Conant, president of Harvard and one of the key leaders of military research during World War II, was the sole and certified candidate for the presidency of the Academy, a quietly conceived rebellion broke out on the floor in favor of Detlev Bronk, long a devoted worker in Academy affairs, and Conant was defeated. The Academy simply announced to the world that Bronk, who had had no part in the rebellion, had been elected to the presidency. Word of this episode was not publicly revealed until seventeen years later.[19]

The Foreign Office of the Academy, which is, in many respects, the State Department of American science, is headed by Harrison Brown, professor of geochemistry and of science and government at Caltech. Brown, an indefatigable worker who has concentrated on bringing the blessings of science and technology to the underdeveloped lands, divides his time between Caltech and his Academy duties—usually spending two weeks per month on each. (He points out that it takes only six hours, door-to-door, from his home in Pasadena, California, to his apartment in downtown Washington.) Brown's chief of staff is an alumnus of the West Coast office of the Central Intelligence Agency, a young man named Murray Todd. Brown explains that he and Todd became acquainted when Todd would debrief him following trips abroad. Brown explains that "I have a protection problem with the kids that we send to Iron Curtain countries [under Academy-sponsored exchange programs]. Todd knows the agency and he can tell them to leave them alone." At least one other "intelligence community" alumnus also works in Brown's office. Brown insists that they have severed all ties with their former employers, and that they were hired for their competence, though, as it has turned out, he says, their intelligence backgrounds have proved useful.

The most influential of the government's science advisory bodies is the President's Science Advisory Committee (PSAC), an eighteen-member group that by custom is chaired by the President's Special Assistant for Science and Technology. But whatever its standing in the hierarchy of influence, its members as often as not go about muttering that the federal establishment is impervious to scientific sense. PSAC is a post-Sputnik outgrowth of a Science Advisory Committee established by President Truman in 1951. With one brief exception, the chairmanship of both bodies has always been filled by alumni of the World War II atomic bomb laboratory at Los Alamos or the radar laboratory—the Radiation Laboratory—at MIT. Academy members have always predominated on PSAC and its predecessor. Until recently, approximately one-third of the members came from Harvard or MIT. Now, however, Cambridge, Massachusetts, is down to one or two members on PSAC, a decline which reflects the powerful presence of Lyndon Johnson in the politics of science.

In 1964, a study found that 29 of the 41 persons who at one time or another held PSAC appointments had received their graduate academic training at seven universities: the University of California, Caltech, Chicago, Columbia, Harvard, MIT, and Princeton. Of the 41, a total of 28 were employed by universities: California (2), Caltech (4), Chicago (1), Columbia (2), Cornell (2), Duke (1), Harvard (4), Illinois (2), Johns Hopkins (1), MIT (5), Pennsylvania (1), Princeton (2), and Stanford (1). The so-called hard sciences (physics, chemistry, mathematics, and engineering) accounted for 77 percent of the total membership; the rest were from the so-called life sciences or other fields, but none was from the behavioral or social sciences—which was still the case in 1967. Finally, it was found that the "hard scientists" were more likely to endure as PSAC members, while their "life scientist" colleagues came and went.[20] Second only to Harvard and Yale's domination of Supreme Court appointments, it can be said that a few universities dominate scientific advice for government.

The prevalence of university-based scientists in the government's science advisory councils is a source of no little concern to scientists who are employed in the government's own laboratories or in industrial laboratories. Thus, Alvin Weinberg, director of the Atomic Energy Commission's Oak Ridge National Laboratory, notes that "even the professor of purest intent must be in some measure loyal to the Estate which he represents. As a result, government scientific advisory circles tend to be preoccupied with science at the universities, rather than with science in industry or in government laboratories; the whole structure and cast of thinking is geared to the problem of university science, and the limitations of the university as an instrument of government are overlooked. It would not be a great exaggeration to describe the advisory apparatus . . . as a lobby for the scientific university." [21]

For proper savoring of this politics, before and during Johnson's presidency, it is vitally important to be acquainted with two closely related administrative techniques that have been fundamental to the science-government relationship since World War II. These are the so-called peer and project systems. They are to pure science what free enterprise is to capitalism; revered, useful, convenient for polemical purposes, and often irrelevant to the real ways of the world.

The peer system, which was developed to an advanced state by private philanthropic foundations before World War II, is based on the premise that scientists themselves are best qualified for certifying the most promising lines of inquiry and the researchers most qualified to pursue them. In practice, this means that when a scientist applies for money for research, opinions are sought from other scientists who are acquainted with his field. They are asked to advise on whether his proposal is soundly conceived, financially realistic, important to the advance of knowledge, and, finally, whether he is qualified to do the work. Since it is doubtful that insurance salesmen can outperform molecular biologists in assessing the quality of proposals for research in molecular biology, the peer system is generally considered to be fundamentally sound in principle. Where it does come in for criticism is on the grounds that it encourages orthodoxy, permits back scratching, and discriminates against the young. But these are criticisms of its operation, not the fundamental premise on which it is founded. The peer system is still well regarded.

The same, however, cannot be said of its longtime partner, the project system, which is based on the fiction that when the government supports a research project in an academic institution, it is supporting the project and not the institution. Fictional though it may be, this interpretation of the intent and effect of government support for research had great political utility in two respects: It helped to open and maintain channels of support for federal aid to science in the universities many years before the Congress was willing to provide direct aid to higher education, and it helped to ward off political pressures for spreading the wealth. Thus, in 1957, the director of the National Science Foundation could tell a Senate subcommittee, "We do not really give the money to the university to use, but we give the money to the university for a scientist for a special problem." [22] When congressmen protested that a few institutions predominated in receipts of the Foundation's money, the director assured them that science, not the institution, was the only beneficiary of this situation. In recent years, however, the project system has been sagging of its own weight. University administrators have had their fill of scientist faculty members whose loyalties run to those who control the coffers in Washington; the have-nots, in smaller institutions as well as in the non-science departments of the great universities, are resentful of the affluence commanded by the prima donnas of scientific research. (W. H. Auden, in "The Dyer's Hand," wrote, "When I find myself in the company of scientists, I feel like a shabby curate who has strayed by mistake into a drawing room full of dukes." [23]) Consequently, while the direct beneficiaries of the project system still depict it as vital to the progress of science, today we find that the elders of the scientific community—that is, administrators who have graduated from the laboratory bench—are expressing grave doubts. Thus, Frederick Seitz, president of the National Academy of Sciences, who was one of the pioneers of solid state physics, recently stated, "In my opinion, the greatest defects of the project system center, first, in the inequities of the selective support given to the sciences and the humanities—inequities that rest on quite basic attitudes prevalent in our society rather than upon skillful political intrigue by the scientist; second, on the circumstance that the project system has tended to denigrate the highly important role of the university administration in establishing policy; and third, on the fact that the system has made it possible for some faculty members in the university community to become very far removed from the life of the undergraduate student." [24]

In general, when it comes to distributing federal largesse to the scientific community, the dominant theme is one of functional disarray. At the White House level there is the Office of Science and Technology (OST), directed by the same scientist who chairs PSAC and who advises the President, the Special Assistant for Science and Technology. OST says it coordinates, though it does not dictate, in conjunction with a subcabinet of officials from government research organizations known as the Federal Council for Science and Technology; the council is also chaired by the President's science adviser. But there is reason to doubt that it gets very far even in coordinating.

It claims that it is doing as well as might be expected in coordinating something that does not lend itself to coordination. But doubts prevail. Thus, in 1966, following a study of federally financed research programs, Representative Henry S. Reuss (D-Wis.) concluded, "The federal government now spends nearly $4 billion annually on research and development in its own laboratories, but it does not know exactly how many laboratories it has, where they are, what kinds of people work in them, or what work they are doing." [25] Reuss is not the most knowledgeable or revered student of science and government affairs, but there is some substance to his allegation. Though he was addressing himself to the government's own laboratories, the less-than-organized state of affairs that he described is prevalent throughout the scientific community and far from accidental. The statesmen of science have violently opposed various schemes that would tidy up their institutional affairs by putting most, or large portions, of scientific research under the roof of one government agency. They like diversity, for not only does it provide a good deal of maneuvering room, but it also assures various opportunities for relief if the congressional seniority system puts the fate of a particular agency into the hands of a congressman who is opposed to big spending for science. The support of basic research is a principal *raison d'être* of the National Science Foundation but the Department of Defense exceeded NSF in this area, $293 million to $207 million in 1966.[26] Research into atomic matters is the principal responsibility of the Atomic Energy Commission, but it is difficult to press nuclear physics into agency boundaries, and money for nuclear research goes forth from NSF, the National Aeronautics and Space

Administration, the Office of Naval Research, and several other agencies. Some twenty federal agencies collaborate in supporting oceanographic research. The great bulk of money for medical research, some 40 percent of the national total, comes from the National Institutes of Health (NIH), but a good deal of the research it supports is indistinguishable from work supported by NSF, the Defense Department, or the Air Force Office of Scientific Research.

As for the recipients of the money, we find that a handful of universities receive most of the money that the federal government provides for research and development in institutions of higher learning. In fiscal 1964, out of some 2100 colleges and universities in the United States (ranging from obscure junior colleges to the well-known giants of the academic scene), ten institutions received 38 percent of the federal funds for university-conducted R and D: the University of California, MIT, Columbia, the University of Michigan, Harvard, Illinois, Stanford, Chicago, Minnesota, and Cornell. Ninety percent of the money was received by one hundred universities. The fact that science and technology tends to cluster, or perhaps that government policies encourage them to cluster, is shown by the fact that institutions in California, Massachusetts, and New York received 48.8 percent of federal funds for university-conducted R and D. A bit more than 76 percent of the funds was received by institutions in these three plus seven other states: Illinois, New Mexico, Maryland, Pennsylvania, Michigan, New Jersey, and Ohio.[27]

The great bulk of the money not only goes to a handful of institutions, but also comprises a substantial portion of the total revenues of these institutions. It is necessary, however, to interpret the amounts with caution, because some of the most striking figures often are a product of an administrative peculiarity of the university-government relationship. Since federal agencies view the academic world as a useful managerial compromise between profit-seeking industry and government ownership and operation, they often entrust the operation of major research facilities to university management. Thus, Johns Hopkins University operates the $50-million-a-year Applied Physics Laboratory under contract to the United States Navy, but the Laboratory is some twenty miles from Hopkins' main campus and has almost nothing to do with Hopkins' own education and research program; rather it does electronics work for the Navy. It would be incorrect to conclude that Hopkins is richer by $50 million a year because of its contract with the Navy, but, like other universities that operate laboratories for the government (Caltech, for example, runs the $240-million-a-year Jet Propulsion Laboratory, under contract to the National Aeronautics and Space Administration), it gets some luster and wealth for its effort. The trustees can take pride in serving their government, but more tangible is the "management fee" that the government pays to the organizations that run these contract laboratories. In the case of Hopkins, this came out to slightly over $1 million in 1966.[28] This may seem like a small sum in relation to the annual operating costs of a major university, but it is equivalent to the annual return on $25 million in endowment invested at 4 percent. What is in it for the government? Hopefully, competent management, the use of the university's prestige for hiring purposes, and a simple way around the restrictive Civil Service pay scales.* Technically, these contract laboratories are nongovernmental, and they need not abide by the government's own salary standards. In the case of MIT, another great government contractor, federal funds, by one reckoning, account for some 84 percent of the institution's total budget. At Princeton, they amount to 47 percent; Columbia, 42 percent; Brown, 35 percent; and Chicago, 28 percent.[29] In all cases, however, the universities that rank as large off-campus contractors for the government are also among the major recipients of federal funds for on-campus research, and, in some instances, have taken steps to intermingle their own campus-based research and training programs with the work that is carried on in the contract laboratories.

Further complicating the administrative picture is a series of varying practices on the part of the money-dispensing agencies. NSF provides money for research; it does not operate laboratories or conduct research. NIH grants money for research in nongovernment laboratories and also runs its own laboratories, spending, in 1964, $539 million for the former, and $109 million for the latter.[30] † The Atomic Energy Com-

---

* The recipients of these fees regularly contend that (1) the sum awarded is simply to help meet the costs of administering a large and complex research operation, and (2) that the federal agencies are niggardly with their management fees. The bookkeeping in these matters is inscrutable, but among outsiders, as well as among a few insiders, it is commonly contended that "management fees" are nothing more than euphemistically termed profits for supposedly nonprofit contractors. In the case of the Johns Hopkins Applied Physics Laboratory, the management there explains, any unexpended balance from the costs of managing the laboratory goes into a contingency fund that is available only for laboratory activities.

† NIH actually spent a great deal more than this, its total budget for the year being $974 million; the balance went into construction, training grants, and other programs.

mission owns laboratories but contracts their operation to nongovernment organizations; it also provides funds for research in university and industry-owned and operated laboratories, but with one minor exception, no research is performed by scientists directly in its employ. The Defense Department owns and operates some laboratories, contracts for others to be operated by nongovernment organizations, and also provides funds for research in university and industry laboratories. Some agencies place their granting decisions largely in the hands of nongovernment scientists who study and evaluate the applications for financial support. For this purpose, NIH and NSF draw upon the advice of thousands of members of the scientific community. Other agencies rely on scientists in government employ. All draw heavily upon the nongovernment scientific community to provide high-level counsel on the scope and quality of their research programs, but the presence of a prestigious advisory committee can just as often be for window dressing as for serious guidance.

The multitude of federal agencies and the value of funds they dispatch to the scientific community has made Washington the administrative and banking center of American science, inspiring the remark that the air routes to the capital are now so crowded with scientists that the airline magazine racks may soon include the *Physical Review* and *Chemical Abstracts* along with the usual periodical fare. The scientists come to Washington for PSAC's monthly sessions, which generally are concerned with broad policy matters; to review the piles of money requests that regularly arrive at the granting agencies; to testify before the growing number of congressional committees concerned with science and technology; to politick at the Cosmos Club; to sniff out NSF's inclination toward continuing a costly project.

Inevitably, there is an immense amount of overlapping in the roles occupied by the cast of characters. Those scientists who find themselves or their institutions in the mainstream of federal support tend to regard the system as logical, economical, functional and even inevitable. The best scientists deserve the most support, they argue, and when the decisions are to be made on whom to support, it is natural that the best people should be consulted. If this suggests the existence of conflicts of interest, the supporters of the system point out that no one may pass on his own applications or even on applications from his own institution. A consultant simply absents himself from the room when such applications come up for consideration, and this is supposed to dispose of any possibility that the matter will be decided on any grounds but intrinsic merit. Underlying the confidence in the system is the belief that the spirit of objectivity traditionally associated with the actual conduct of research can readily be transferred to the committee room where administrative and policy decisions are made. Those scientists who are outside the mainstream of support naturally have their suspicions. They note, for example, that the University of California, which is the largest single academic recipient of federal research funds, also provides the largest number of members for federal review panels. To which the Californians can reply that they possess more Nobel laureates (12) than any other institution, more members of the National Academy of Sciences (97), and in 1966, received more of the highly coveted Guggenheim Fellowships (50) than any other university.[31] The Californians lump together their nine statewide campuses for these compilations, but the Berkeley campus alone contains 9 of the Nobel laureates, 51 Academy members, and in 1966 was cited by the American Council on Education as the "best balanced distinguished university" in the nation.[32] Not surprisingly, it is first among the country's universities in NSF grants as well as in postdoctoral grants in humanities and social sciences of the American Council of Learned Societies. In any case, according to one study, the ten universities that received 38 percent of federal research funds provided 36.8 percent of the advisory panel membership and accounted for 54.2 percent of Academy membership.[33] Is it that the rich undeservedly get richer, or that the rich deserve their riches? Scientists who have become affluent under this system credit it for the indisputable quality and productivity of contemporary American science. Scientists clamoring to get into the system dispute the relationship between ability to obtain federal support and the ability to do good science. They point out that many of the luminaries at major universities made their scientific reputations in the minor leagues, and then moved on.

The sense of injustice that prevails in many smaller institutions is revealed in the remarks of Harvey R. Fraser, president of the South Dakota School of Mines and Technology. "For several years, from 1960 to 1965, we had a very energetic, capable, research minded Ph.D. on our staff. He submitted numerous proposals [to government agencies] but was only modestly successful in obtaining grants. Two years ago, this man resigned and transferred to a large university. In one year, he had more grants and more research than he could handle. We had and do have outstanding facilities in his area. His capabilities for research did not suddenly generate the day he left a small school for a big one." [34] *

Laymen, upon first acquaintance with the scientific community's ways of doing business, often are appalled and outraged. That scientists should be both recipient and principal adviser to their public patron runs counter to popular notions of man's capacity for integrity in the face of temptation. But as was noted earlier, science is unlike any other activity, and the scientific community is unlike any other organization.

These deviations from tidiness and traditional governmental procedures reflect the paradox with which this chapter began: appearances to the contrary, there is in fact no clear-cut Scientific Establishment in the United States. There is a scientific community with its elite and its lower classes, with a functional division of labor, and with organizations that both structure the community and define its relationship to the nation at large. As between the community and the nation as well as among the scientific organizations themselves, there is a vital and continuous struggle for money and the power to grow. This struggle in all of its complexity is the politics of science.

Now, let us look more deeply into the affairs of the scientific community, with particular emphasis on the folkways of its inhabitants.

NOTES

1. *The Logic of Liberty* (Chicago: University of Chicago Press, 1951), p. 89.
2. *13th Annual Report,* National Science Foundation, 1963, p. xix.
3. *Chemistry: Opportunities and Needs,* National Academy of Sciences–National Research Council, 1965, p. 151.
4. *Scientific and Technical Manpower Resources, Summary Information on*

* This tale of difficulty, voiced before a Senate investigating committee, was informally looked into by a White House science aide, who concluded that what did change was the scientist's teaching load. In South Dakota, the aide concluded, the scientist was so loaded with teaching duties that the federal granting agencies did not believe he would have sufficient time for a sizable research program. At the large university to which he transferred, he held relatively light teaching duties. All of which raises a series of new issues—the effects of research on teaching. Though it might be difficult to trace the sources, it is reasonable to assume that federal subsidy of one sort or another was the underpinning of the large university's ability to permit ample time for research.

Employment, Characteristics, Supply, and Training, National Science Foundation, 1964, NSF 64-28, pp. 18–20.

5. *13th Annual Report,* National Science Foundation, 1963, p. 132.
6. Robert C. Wood, "The Rise of an Apolitical Elite," *Scientists and National Policy Making,* Robert Gilpin and Christopher Wright, eds. (New York: Columbia University Press, 1964), p. 48.
7. Vannevar Bush, *Science, The Endless Frontier* (1945; reprinted by National Science Foundation, 1960), p. 86.
8. *Federal Funds for Research, Development, and Other Scientific Activities,* National Science Foundation, 1964, NSF 64-11, p. 52.
9. *Scientific and Technical Manpower Resources,* pp. 4–5.
10. *Ibid.*
11. *Basic Research and National Goals,* report to the Committee on Science and Astronautics, House of Representatives, by the National Academy of Sciences, 1965, p. 308.
12. "Some Problem Areas in the Relationships Between Government and the Universities," address to National Academy of Sciences, October 13, 1964, *News Report,* National Academy of Science–National Research Council, November-December 1964.
13. Mark Twain, *The Adventures of Huckleberry Finn* (New York: Grosset & Dunlap, 1948), pp. 153–54.
14. *Basic Research and National Goals,* pp. 308–9.
15. *Ibid.,* p. 311.
16. *Scientific and Technical Societies of the United States and Canada,* Publication 900, National Academy of Sciences–National Research Council, 1961.
17. "Analysis of Ages of Membership of the Academy, as of July 1, 1963," mimeographed paper, National Academy of Sciences, January 1964.
18. D. S. Greenberg, "The National Academy of Sciences: Profile of an Institution," three-part series, *Science,* April 14, 21, 28, 1967.
19. *Ibid.,* April 21, 1967, p. 360.
20. Carl W. Fischer, Jr., "Scientists and Statesmen, A Profile of the Organization and Functions of The President's Science Advisory Committee," paper prepared for Science and Public Policy Seminar, Graduate School of Public Administration, Harvard University, 1964.
21. Alvin W. Weinberg, "The New Estate," *Yale Scientific Magazine,* October 1963, p. 16.
22. *Senate Independent Offices Appropriations, Hearings,* 1958, p. 276.
23. *The Dyer's Hand and Other Essays* (New York: Random House, Inc., 1962), p. 81.
24. Frederick Seitz, "The University: Independent Institution or Federal Satellite?", *Science and the University,* Boyd R. Keenan, ed. (New York: Columbia University Press, 1966), p. 155.
25. Research and Technical Programs Subcommittee, House Government Operations Committee, press release, November 14, 1966.
26. *The Budget of the United States, Year Ending June 30, 1966,* p. 446.
27. *Impact of Federal Research and Development Programs,* Study No. 6, report of the Select Committee on Government Research, House of Representatives, 1964, pp. 33–34.
28. *House Appropriations Committee, Hearings for 1968,* Department of Defense, Part 3, p. 352.
29. *Impact of Federal Research and Development Programs,* Study No. 6, pp. 36–37.

30. *Basic Data Relating to the National Institutes of Health,* Department of Health, Education and Welfare, 1965, p. 19.
31. *Facts and Figures about the University of California* (University Relations Office, November 1966).
32. *An Assessment of Quality in Graduate Education,* American Council on Education, 1966, p. 107.
33. *Administration of Research and Development Grants,* Study No. 1, report of the Select Committee on Government Research, House of Representatives, 1964.
34. *Senate Committee on Government Operations, Subcommittee on Government Research, Hearings,* July 27, 1966.

II
Chauvinism, Xenophobia, and Evangelism

By the worldly standards of public life, all scholars in their work are oddly virtuous. They do not make wild claims, they do not cheat, they do not try to persuade at any cost, they appeal neither to prejudice nor to authority, they are often frank about their ignorance, their disputes are fairly decorous, they do not confuse what is being argued with race, politics, sex or age, they listen patiently to the young and to the old who both know everything. These are the general virtues of scholarship, and they are peculiarly the virtues of science.

J. BRONOWSKI, in *Science and Human Values,* 1956

Perhaps the time has come for them [scientists] to wonder about why they sometimes jar the nerves and try the patience of non-scientists.

The humanist who looks at science from the viewpoint of his own endeavors is bound to be impressed, first of all, by its startling lack of insight into itself. Scientists seem able to go about their business in a state of indifference to, if not ignorance of, anything but the going, currently acceptable doctrines of their several disciplines. . . . The only thing wrong with scientists is that they don't understand science. They don't know where their own institutions came from, what forces shaped and are still shaping them, and they are wedded to an anti-historical way of thinking which threatens to deter them from ever finding out.

ERIC LARRABEE, in "Science and the Common Reader," *Commentary,* June 1966

To proceed with our inquiry into the politics of pure science it is essential to take note and be mindful of an enduring predicament of pure science and a complex set of behavioral characteristics that are evoked by the predicament.

The predicament is that pure science is neither self-explanatory nor self-supporting. Its affiliate, technology, is both, and consequently has easily acquired a mass constituency, something on the order of the mass constituency that formerly gave allegiance to religion. But pure science, as distinguished from technology, remains far beyond the comprehension of the general public; in fact, contemporary pure science is so specialized, so esoteric, that few scientists today possess any significant understanding of research outside their own field. That is one side of the predicament. The other is that while basic research grows ever more costly, it has nothing to sell in the conventional marketplace. Technological research is explicitly directed toward the creation of tangible, purposeful, salable goods. But the only tangible consequences of pure research are scientific papers for the consumption of other scientists. Applied and developmental research are visibly related to the fulfillment of worldly, utilitarian objectives, such as the desalting of sea water, faster, safer air transportation, improved communications. There is a popular interest in and demand for the fruits of this research. The public can literally touch the consequences of such research, and economists can more or less measure their value. But what is the value of and where is the constituency for a typical product of basic research, such as the following randomly selected example: "The synthesis of carbamoyl phosphate required in both arginine and pyrimidine biosynthesis is carried out by a single enzyme in *Escherichia coli*. Opposed effects of pyrimidine nucleotides and of ornithine on the activity of the enzyme ensure a proper supply of carbamoyl phosphate according to the needs of the two biosynthetic sequences." [1] Ultimately, though unpredictably, such findings may have great value for the public that is asked to support science. But, at any given moment, the consumers of such knowledge probably number no more than a few score workers, scattered about the world, who are engaged in similar research.

The knowledge embodied in a scientific paper is neither useful, salable, nor patentable until it is transformed into a technique, a material, a pill, or some other tangible form. Thus, when the public is asked to support science, it is, from its own scientifically illiterate perspective, being asked to support the production of incomprehensible intangibles.

Historically, this twofold predicament has been an enduringly difficult burden on science. Except in extraordinary circumstances, it has pressed in on the scientific community, helping to nurture, in response, the development of a trinity of behavioral characteristics that pervade the scientific community and shape its relations with the outside world. Confronted by lack of understanding, skepticism, and, at times, ill will in their nonscientific surroundings, the inhabitants of the scientific community are, first of all, reverently patriotic toward the methodology and mores of their craft. Secondly, they are anxiously poised to expel intruders who would usurp the name of science or meddle in its internal affairs; and finally, they simply wish the rest of us would convert to science. Let us say that this trinity of characteristics can best be summed up as chauvinism, xenophobia, and evangelism, and then let us proceed to examine them in their effects on the community, and, in turn, on the community's relation with the outside world.

In one way or another, all professions share these characteristics in the sense that they esteem their own role in society, are opposed to outlanders interfering with that role, and wish to win converts, if not to their ranks, at least to a recognition of their place in the world. Whether the scientific community exceeds the norm for devotion to these characteristics defies precise measure; nor do all scientists partake equally of the trinity. As we shall see, among scientists there is distress and contention arising from issues related to these characteristics. Nevertheless, a traveler through the scientific community finds it abounding with the belief that (1) science is a noble, and perhaps the noblest, creation of the human intellect, (2) it is too delicate and valuable to be subjected to lay intrusion, but (3) it is so powerful, beautiful, and useful that an understanding of it must be conveyed to the lay world. Reversing and amending the traditional maxim, many scientists believe that science is too important either to be intruded upon or ignored by laymen. From these beliefs have flowed many consequences in the real world of science and politics.

In the pure scientists' hierarchy of values, the acquisition of knowledge ranks, a priori, above the application of knowledge. "He who makes two blades of grass grow where one grew before is the benefactor of mankind," Professor Henry A. Rowland, a Johns Hopkins physicist, declared in 1879, "but he who obscurely worked to find the laws of such growth is the intellectual superior as well as the greater benefactor of mankind." [2] Eighty-six years later, one of the leading statesmen of science, George B. Kistiakowsky, in effect, uttered the same sentiment: "The point that requires repeated emphasis," he stated, "is that closely defined mission-oriented research [another term for applied or developmental research] has value but, by itself, is insufficient and incapable of developing really new ideas and new principles on which each practical mission will ultimately find itself based. If the social climate and support mechanisms are not such as to encourage the free exploration of new ideas rapidly and effectively, our technology will die on the vine because, in the absence of the results of new, undirected basic research applied work tends to become more and more confined to increasingly expensive refinements of and elaborations of old ideas." [3]

The faith inherent in the theses of Rowland and Kistiakowsky possesses a large and readily perceived beauty, as well as a nugget of truth. Pure research has verifiably made indispensable contributions to technology. The only difficulty is that when the origins of technological developments are subjected to systematic, scholarly analysis, the question of the debt to pure research turns out to be riddled with far more paradoxes, puzzles, and uncertainties than the statesmen of pure science are generally willing to admit. There are, to be sure, striking instances of fundamental, undirected, research—research for the sake of research —leading to the acquisition of knowledge that was then transformed into new, valuable, and powerful technology. Atomic energy is, of course, the most vivid example of this sequence, and, in the postwar years, the recitation of the tale of the atom has become commonplace among the statesmen of pure research in their appeals for public support. But, as often as not, the history of science and technology fails to conform to the pure scientists' tidy model of science as the father of technology. It would be convenient, for example, if a comprehension of thermodynamics had paved the way to the creation of the steam engine, but, if anything, it appears that the steam engine paved the way to a comprehension of thermodynamics—and the inspiration for this effort at comprehension was a desire for a still more efficient steam engine. Which came first, which was of greater importance, the science or the technology? Upon close examination, it becomes clear that these are the wrong questions; that rather than a straight-line sequence from knowledge to utility, there has prevailed an interaction of such incredibly complex and intricate composition that it is rarely possible in examining any artifact or device to sift the science from the technology.* Henrik Bode, of Bell Telephone Laboratories, observes of the eighteenth and nineteenth centuries that ". . . in a certain sense, science was far more indebted to technology than technology to science throughout this period. There were, of course, exceptions, but on balance the

---

* An excellent introduction to this subject is to be found in Richard R. Nelson, "The Economics of Invention: A Survey of the Literature," *Journal of Business,* April 1959.

scientist was in the position of relying on technology, or, more broadly, on the world of practical experience generally, for his tools and much of his information. Technology 'was there first.' For example, the invention of both the telescope and the microscope depended on a flourishing industry in spectacle lenses that already existed. Magnetism was known as an empirical fact, and had been used as a basis for the navigator's compass for centuries before 18th and 19th century physicists got around to studying the phenomenon. Watt's steam engine was invented without the benefit of the Carnot cycle, or Joule's work, and so on." [4]

The ideologists of basic research would have us believe that today the dynamics of the western world's scientific and technologically based culture have radically transformed this process, that the process of moving from knowledge to utility has now been both well established and accelerated.

Several years ago, for example, Glenn T. Seaborg, chairman of the Atomic Energy Commission, declared that science "has now become the basis for the advance of our economy," [5] thereby echoing the litany that proponents of basic research have regularly uttered since the establishment of the science-government partnership at the end of World War II. (At that time, for example, it was stated that "advances in science, when put to practical use, mean more jobs, higher wages, shorter hours, more abundant crops, more leisure for recreation, for study, for learning to live without deadening drudgery. . . ." [6])

Is this really so? Until quite recently, the answer of pure science was generally an unqualified "Yes." And the public, capable of neither understanding nor dismissing the pronouncements of the prestigious statesmen of science, came to accept the thesis that pure knowledge inevitably leads to utilitarian results. However, under skeptical assessment from scholars outside the physical and natural sciences, as well as pressure to justify its ever-increasing demands for public support, the community of science itself has recently been pushed to the realization that the relationship between science and utility is infinitely more complex than it had hitherto been willing to concede.* This is a difficult

* The point was well stated by Milton Harris, vice president of research for the Gillette Company, in testimony before the Senate Small Business Committee in 1963. Harris, who had headed a study of civilian technology for the White House science office, pointed out, "So much has been written and spoken about the beneficial impact of science and technology on our national security and economic growth that too many people have come to accept with religious fervor

realization, and the statesmen of science tend generally to approach it gingerly. For example, Harvey Brooks, an ardent advocate of pure research, treats the problem as follows: "There is now a general acceptance among economists of the importance of technological innovation in economic growth. To an increasing extent such innovation depends upon the results of basic science, although the degree to which this is true is difficult to quantify." [7]

However, outside the community of pure science, there are those who go still further in asserting the tenuousness of the relationship between scientific knowledge and utility. In 1963, for example, the Department of Defense, after nearly two decades and several billion dollars investment in basic research, mainly at universities, undertook a study which it called Project Hindsight.[8] What the Defense Department did was to select twenty weapons that in large part comprise the technological backbone of this nation's military security. (Included were various nuclear warheads, rockets, radar equipment, a navigation satellite, and a naval mine.) Each of these weapons was subjected to careful analysis to determine the origins of the science and technology that made its existence possible. Since the Department had been supporting science and technology to a significant extent only since 1945, the study was limited to identifying only those scientific and technical contributions that had occurred since that year. Hindsight, to the great chagrin of the basic scientific community, concluded: "In the systems which we studied, the contributions from recent undirected research in science [the study's term for basic research] was very small." Defining each discrete, identified contribution to a weapon as an "event," the study concluded that the twenty weapons studied embodied a total of 556 events. Of these, 92 percent came under the heading of technology; the remaining 8 percent were virtually all in the category of applied research, except for *two* which fell under the definition that we have been using for pure science.

Hindsight is, of course, an egregious outrage to the ideology of pure

the concept that research and development constitutes the basic and supporting strength of our society. . . . There is a strong tendency today to equate economic growth with the size and intensity of the nation's research and development activities. . . . Without in any way wishing to downgrade the importance of my chosen profession, I think it is self-evident that many factors are extremely significant in determining the economic growth rate—the tax structure, monetary policies, wage policies and restraints, incentives to capital investment, antitrust issues, to mention just a few." (Subcommittee of the Senate Select Committee on Small Business, *Hearings,* Part 2, June 1963, p. 228.)

science.* And it must be observed that the study acknowledges its own deficiency by noting that it is concerned solely with post-1945 science, that "without the organized body of physical science extant in 1930—classical mechanics, quantum mechanics, relativity, thermodynamics, optics, electromagnetic theory and mathematics—none of our applied science and only a fraction of technology events could have occurred." Moreover, Hindsight pays little attention to the fact that basic research is quite often the training ground for technologists. Nevertheless, no matter how repellent it may be to the ideologists of basic science, Hindsight, at the very minimum, erodes a simplistic myth that they have created and, with great energy, diffused through the postwar period, namely, that basic research is the underpinning of contemporary technological development. In terms of better weapons, the Hindsight study concludes, the Department of Defense has *so far* received very little from its massive postwar investment in basic research.† Or, as it was put another way by one of the most sophisticated students of the relationship between science and technology, J. Herbert Hollomon, a General Electric engineer whom the Kennedy administration appointed to the newly created position of Assistant Secretary of Commerce for Science and Technology: "New technology flows from old technology, not from science. . . . Most technological developments . . . go one step at a

---

* One of the leading statesmen of pure research privately protested the publication in *Science* of a news article that described the Hindsight report. Description of so heretical a thesis, he felt, was tantamount to advocacy of it.

† Hindsight has been criticized on the ground that it fails to distinguish the qualitative values of the "events" under consideration. Thus, the transistor, developed by Bell Laboratories, is counted as a single event, though there is no doubt that most modern weapon systems would be impossible without transistors and other semiconductor devices. It is interesting to note, however, that the transistor was not the result of the serendipity and free-form activity that is so revered by basic science. Rather, as Richard R. Nelson points out, Bell was seeking new amplifying devices and skillfully integrated the scientists' traditional yen for knowledge with its own clearly established objectives. William Shockley, who received the Nobel Prize for his work on the transistor, allows for no dichotomy between basic and applied research. "Frequently, I have been asked if an experiment I have planned is pure or applied science; to me it is more important to know if the experiment will yield new and probably enduring knowledge about nature. If it is likely to yield such knowledge, it is, in my opinion, good fundamental research; and this is more important than whether the motivation is purely esthetic satisfaction on the part of the experimenter on the one hand or the improvement of the stability of a high-power transistor on the other." (Quoted in Richard R. Nelson, "The Link Between Science and Invention: The Case of the Transistor," *The Rate and Direction of Inventive Activity* [Princeton: Princeton University Press, 1962].)

time: Improvement of artifact or process, and then a gradual cumulation of the state-of-the-art rather than a gradual cumulation of the literature of science." Moreover, Hollomon points out, "The support of science is not a *sine qua non* for economic and social development. More often it is not that science produces wealth but that wealth can support science." Citing the remarkable economic growth of Japan, Hollomon observes that "From 1945 to 1960, Japan grew economically at the startling rate of 10 percent a year, yet until recently the support of basic science in Japan was negligible." [9] *

The point that I earnestly strive to make—and which is quite likely to be misunderstood by the ideologists of pure science—is not that basic research is of inconsequential importance to contemporary society; nor is it that basic research has wittingly misrepresented itself in its quest for support. Rather, I strive to make clear that largely in response to the predicament of being neither self-explanatory nor self-supporting, basic research has had an incentive, for purposes of survival and growth, to claim certainty when, at most, it could establish only probability; it has had incentive to ascribe to itself clear-cut economic significance, when, in fact, neither scientists nor economists have anything but a dim understanding of the role that science plays in economic development.†

* The cult of research, pure and otherwise, holds forth research as the solution for many social ills that, in fact, have little or nothing to do with lack of knowledge. Jacques Barzun, in his brilliant, witty, and profoundly important book *Science: The Glorious Entertainment* (New York: Harper and Row, 1964), brings to our attention (p. 42) the letter of a *New York Times* reader commenting on proposals for research on 200-mile-an-hour rail service. "There is no need," says the writer, "for another expensive foundation research project to begin railroad recovery. Any rider can provide expert advice, free of charge, to management.

"When it is cold, the cars should be heated; when it is warm, they should be cooled; where it says 'No Smoking' the conductor should enforce it . . . where there is scenery, it should be discernible through the window . . . where there are rest rooms, there should be sanitation somehow comparable to a well-run stockyard."

† Addressing himself to the question of how much support the federal government should provide for basic research, Harry G. Johnson, professor of economics at the University of Chicago, observes: "Though the importance of the advance of knowledge to improved living standards is difficult to quantify, and the magnitude of the contribution of basic research to the advance of productivity still more obscure, and though both may be easily exaggerated in carelessly formulated argument, there is no disputing that basic research has played a significant part in the growth of the U.S. economy. This fact by itself, however, does not constitute a case for Government support of basic scientific research, though scientists frequently write as if it did. . . . In order to establish a case for Government support, it must be shown that basic research yields a social return

It is not improbable that the French monarchy saw intrinsic validity in the theory of the divine right of kings. When the statesmen of pure science proclaim that pure science is the locomotive of contemporary society, their sincerity is equaled only by their lack of evidence. Nevertheless, they do believe it. And now let us go on to see how this belief affects their perceptions and actions.

In harmony with the belief in the superiority of basic science, the National Academy of Sciences has so limited the admission of applied scientists and engineers that in 1965, engineers established their own honorary institution, the National Academy of Engineering. To the lay world, Jonas Salk, developer of the polio vaccine, is a scientist. But the pure science community slices its distinctions finely. Salk's vaccine, based on killed virus, was derived from tissue culture techniques for which three basic researchers, John Enders, Thomas Weller, and Frederick Robbins, were awarded the Nobel Prize in 1954. Though proposed for membership in the National Academy of Sciences, Salk was never accepted. Why? The Academy offers no explanation for its membership decisions, but the late Thomas M. Rivers, an Academy member who headed the hospital of the Rockefeller Institute, was quoted as citing "original work" as the key to Academy membership—and explaining: "Now I'm not saying that Jonas wasn't a damned good man, but there have been killed vaccines before. Lots of them." [10]

The basic scientists' hierarchy of values has so permeated higher education during the postwar period that even many basic scientists have expressed concern about the effects on scientific and technical education in general, and on applied science and engineering in particular. Edward Teller, a basic nuclear scientist who has become an apostle of the need to improve the quality and status of applied science, has observed, "In our educational institutions applied science may almost be described as a 'no-man's land.'" [11] Similarly, Alvin M. Weinberg, who came from the ranks of basic research to become direc-

---

over its cost that exceeds the return on alternative types of investment of resources." Johnson concludes that at present, however, information is lacking to establish a case one way or the other. He also observes that "insistence on the obligation of society to support the pursuit of scientific knowledge for its own sake differs little from the historically earlier insistence on the obligation of society to support the pursuit of religious truth, an obligation recompensed by a similarly unspecified and problematic payoff in the distant future." ("Federal Support of Basic Research," *Basic Research and National Goals,* a report by the National Academy of Sciences, 1965, pp. 127–41.)

tor of the Oak Ridge National Laboratory, has chided his fellow scientists for their supercilious attitudes toward applied research, noting that "most of the prestige and emphasis in the university goes to basic science. The best scientific minds go into basic, not applied, science; and the social hierarchy of science, reflecting the discipline-orientation of the university as much as it does the intrinsic logic of the situation, places pure science above the interdisciplinary applied science." [12] And, finally, Philip H. Abelson, editor of *Science* and director of the Carnegie Geophysical Laboratory, has warned that the basic scientists' chauvinism is neither realistic nor beneficial to society. Pointing out that "professors have looked down on nonuniversity research, have regarded its practitioners as inferiors, and have attempted to curtail their activities," Abelson noted, "Most university science graduates must eventually find employment in nonacademic posts. When they do they accept for themselves what they have been taught is a second-class status. As a result they can have deep loyalty neither to their alma mater nor to their employer." [13]

The basic scientists' reverence for their profession has caused some normally rationalistic scientists to offer up quasi-mystical propositions in justification of their quest for public support.

"Basic research," writes Harvey Brooks, "is recognized as one of the characteristic expressions of the highest aspirations of modern man. It bears much the same relation to contemporary civilization that the great artistic and philosophical creations of the Greeks did to theirs, or the great cathedrals did to medieval Europe. In a certain sense it not only serves the purposes of our society but *is* one of the purposes of our society." [14] *

The concept of science as a *purpose* of society is bold and brilliant, and, it must be recognized, is not inconsistent with the intent stated in

* The cathedral metaphor occurs repeatedly in the public pronouncements of the statesmen of science, as, for example, in the words of Philip Handler, chairman of the biochemistry department at Duke University, chairman of the National Science Board, and a member of the President's Science Advisory Committee: "The edifice which is being created by science . . . is fully comparable to the cathedrals of the Middle Ages or to the art of the Renaissance. . . ." (Remarks quoted in press release titled "Science or Technology, They Need Each Other," Office of Information Services, Duke University, Feb. 5, 1967.) That the building of pyramids and cathedrals exacted a monstrous toll from the masses that were supposedly elevated by these edifices is never discussed.

the opening words of the National Science Foundation Act of 1950—
"An act to promote the progress of science." But the concept of science
as the contemporary counterpart of the medieval cathedral unfortunately clashes with a troublesome reality, namely, the increasingly
esoteric nature of today's science. The masses could see and, presumably, enjoy and be spiritually elevated by the cathedrals that absorbed
their labor and wealth. But modern science not only is inscrutable to
the masses, but, in many of its manifestations, has become progressively
dissociated from humankind. As J. Robert Oppenheimer pointed out in
an address to the bicentennial observation of the Smithsonian Institution, "I have sometimes asked myself when a discovery in science
would have a large effect on beliefs which are not, and may perhaps
never be, a part of science. It has seemed clear that unless the discoveries could be made intelligible they would hardly revolutionize
human attitudes. But it has also seemed likely that unless they seemed
relevant to some movement of the human spirit characteristic of the
day, they would hardly move the human heart or deflect the philoso-
pher's pen. . . . Five centuries ago the errors that physics and astronomy
and mathematics were beginning to reveal were errors common to the
thought, the doctrine, the very form and hope of European culture.
When they were revealed, the thought of Europe was altered. The errors
that relativity and quantum theory have corrected [in this century] were
physicists' errors . . . limited to a very small part of mankind." [15]

Or, consider this expression of concern by Weinberg as to the direction and values of contemporary science: "In making our choices for
allocating scientific resources we should remember the experiences of
other civilizations. Those cultures which have devoted too much of their
talent to monuments which had nothing to do with the real issue of
human well-being have usually fallen on bad days: history tells us that
the French Revolution was the bitter fruit of Versailles, and that the
Roman Colosseum helped not at all in staving off the barbarians . . .
we must not allow ourselves, by short-sighted seeking after fragile
monuments of Big Science, to be diverted from our real purpose, which
is the enriching and broadening of human life." [16]

When a handful of particle physicists, at the expense of several hundreds of millions of dollars per year in public funds, explore the "particle zoo," are they fulfilling a purpose of society? Or are they merely
pursuing their own curiosity in virtually total disengagement from the

society that supports them? These are questions for which the statesmen of science have no simple answers.

The affection that scientists feel for their professions often is extraordinary. "My chief enjoyment and sole employment throughout life has been scientific work," wrote the ailing Charles Darwin, "and the excitement for such work makes me for the time forget, or drives away, my daily discomfort." [17] Or consider the words of George W. Beadle, the Nobel laureate biologist who is president of the University of Chicago: "It is difficult to explain to one who has never experienced it, the incomparable thrill, excitement, and satisfaction of original discovery. The memory of my own first discovery in science remains remarkably vivid. It was a small finding—the identification of a gene in Indian corn. . . . Even so, I rode the clouds for weeks." [18]

Or, consider the blend of chauvinism and evangelism in the words of Glenn T. Seaborg, the Nobel laureate chemist who is chairman of the Atomic Energy Commission: "The message is clear—science and technology are the wave of the future. . . . The educated man of today and tomorrow can no more ignore science than his predecessors of the Middle Ages could ignore the Christian church or the feudal system." [19] Whether the scientifically illiterate are to be excluded from the ranks of the "educated" is a matter of definition, but that it is becoming increasingly difficult to seek education without being exposed to science is a fact of contemporary America. Never has any profession so assiduously organized and labored to make its stuff known to the masses. In 1966, an admittedly incomplete survey reported sixty-two organizations in the United States developing new courses of instruction in science and mathematics. These ranged from the Commission on Science Education of the American Association for the Advancement of Science to a project at Wesleyan University, titled "Writing and Testing a Textbook and a Teacher's Commentary for a Tenth Year High School Course in Modern Coordinate Geometry." [20] Nor is this evangelism confined to the United States. The same study found twenty-nine similar organizations, international, national, and local, working abroad. These ranged from the African Mathematics Program to Turkey's Mobile Units for Science Teaching. These activities are admirable, useful, and long needed. But it is worth noting that nothing approaching this missionary effort has been undertaken by the devotees of law, art, economics, or philosophy. There is a good reason, of course: the methods and products of science and technology are more "useful," and nations realizing this seek to train their youth in these fields.* But in most cases it is the missionary zeal of scientists that has brought nations to this realization. Further, it is not just "scientists," but a particular brand of scientists—pure scientists, driven by evangelistic fervor —who have tended to dominate the curriculum reform movement. As Weinberg points out, the curriculum reforms, starting in the high schools and then spreading to the lower and upper levels of education, "have been instigated by the university, and they certainly reflect the intellectual spirit of the university. With certain aims of the curriculum reform, one can have no quarrel. The new curricula try hard to be interesting, and in this I think they succeed; also they demand more effort and present more challenge than the old. But, insofar as the new curricula have been captured by university scientists and mathematicians of narrowly puristic outlook, insofar as the curricula reflect deplorable fragmentation and abstraction, especially of mathematics, insofar as the curricula deny science as codification in favor of science as search, I consider them to be dangerous. . . . The professional purists, representing the spirit of the fragmented, research-oriented university, got hold of curriculum reform, and by their diligence and aggressiveness, created puristic monsters." [21] †

---

* Humanists sometimes look with envy upon the curriculum reform movement in the sciences. "The science programs of the secondary schools have never been so imaginative, so effective, and so sophisticated as they are today," stated Chairman Barnaby C. Keeney in the First Annual Report of the National Endowment for the Humanities in 1966. "Since the Second World War, there has been a revolution in the teaching of science at all levels. No such phenomenon can be observed in the humanities," he concluded.

† One source of great chagrin and bewilderment to the physicists is that while they have pioneered in devising new courses of instruction in physics for use in high schools and colleges, the students are not flocking to the study of physics. In the academic year 1960-61, a total of 254,215 undergraduate degrees were granted in the U. S.; of these, 5293 were in physics. By 1964-65, the total had risen to 314,000, but degrees to physics majors had risen only to 5517. The physicists project that by 1970, when an estimated 450,000 degrees will be awarded, the number of physics majors among them will have actually declined to approximately 4500. Why? One answer they offer is that "In some schools the introduction of the PSSC (Physical Science Study Committee) course has reduced the number of prospective physics majors." (*Physics Manpower, 1966, Education and Employment Statistics,* American Institute of Physics, AIP Pub. No. R-196, p. 36.) Of course, the question of why students study this rather than that is a complex one, bound up with course requirements, employment prospects, the availability and quality of instruction, social and political values, and so forth. But it is also quite likely that the physicists have simply made it too tough for the kids.

In the biological sciences, the molecular biologists, who are the purest of the pure biologists, have seen to it that their specialty is heavily emphasized in the many new curricula that have been prepared in recent years for high school and college instruction in biology. With what results? Following is the testimony of a professor of anatomy at the University of Michigan Medical School: "Recently, our department gave an examination to our entering freshmen medical students prior to the formal course in order to assess their background for microscopic anatomy. . . . Over 80 percent of our entering students were knowledgeable about the structure of DNA, the triplet coding system as well as many other features of cellular biology. On the other hand, less than 15% had any appreciation of tissue composition and less than 2% were aware of organ structure and function. . . . It is indeed curious that a 'pre-med' student is thoroughly familiar with the esoterics of molecular biology while he has little or no appreciation for tissues and organs and the essential processes involved in their function." [22]

Within the scientific community, the evangelical spirit is intense, but even more intense is the xenophobic aversion that the community feels toward heathens who presume to speak in the name of science. Thus, twenty years ago, when Immanuel Velikovsky, a psychoanalyst, propounded cosmological theories wholly at variance with generally held scientific theories, the astronomical fraternity did not simply write him off as a pseudoscientific crank. Rather, it went to extraordinary lengths in an unsuccessful effort to suppress publication of his book, *Worlds in Collision*. Harlow Shapley, then director of the Harvard College Observatory, declared, "if Dr. Velikovsky is right, the rest of us are crazy." [23] Some scientists attempted to boycott Velikovsky's publisher, Macmillan. Several denounced the book, though admitting they had not read it.[24]

The xenophobia of scientists is intimately related to their reverence for the methodology and formalities of science. Nature reveals its secrets grudgingly, and the presumption, backed by ample evidence, it must be recognized, is that these secrets are likely to be obtained only by those who have been formally initiated into the ranks of science and who employ the traditional techniques of science. Thus, when the monk Gregor Mendel discerned the laws that govern the passage of hereditary characteristics, his contemporaries in the established ranks of science refused to pay attention. As Bernard Barber points out, "Mendel sent his paper to one of the distinguished botanists of his time, Carl von Nägeli of Munich. Von Nägeli resisted Mendel's theories for a number of reasons: because his own substantive theories about inheritance were different and because he was unsympathetic to Mendel's use of mathematics, but also because he looked down, from his position of authority, upon the unimportant monk from Brünn." [25] In 1965, when a panel of scientists convened by the National Academy of Sciences reported on the state of knowledge in weather modification, W. E. Howell, a commercial rainmaker, protested that the "panel didn't take a good look at commercial cloud-seeding results before publishing its preliminary report. . . ." Howell stated, "the fact is that our efforts to get our results published and seriously discussed were met with rebuffs that plainly told us our betters thought we were usurping the rostrum of science for our own money-grabbing and unscientific ends." He added that some of the commercial reports were filed with the Weather Bureau library, and thus were available for the panel's consideration. To which a panel member, James E. McDonald of the Institute of Atmospheric Physics at the University of Arizona, commented that ". . . it had never been the panel's mission to undertake a laborious analysis of commercial reports lying wholly outside the open scientific literature. . . . I cannot agree with any suggestions that filing such reports in the Bureau library comes anywhere near publishing results in carefully checked and fully documented form in the open scientific literature." [26]

A scientist reviewing a book on extrasensory perception writes that the author has "little notion of what experiments are, and less liking for the methodological niceties of scientific research." The author protests, and the reviewer replies: "I am afraid that it would require a few years' training in a first-rate department of experimental psychology to give him an appreciation of what the term 'experiment' properly means; I could not hope to do so in a few sentences." [27]

The scientists' reverence for their profession and their lofty vision of its place in the world make them acutely concerned about the manner in which the lay world respects and comprehends science. Thus Arthur Compton, whose chauvinism is extreme but not unique, states that when James B. Conant, the chemist who was president of Harvard, advised Roosevelt to proceed with the development of the atomic bomb, it was not simply Conant's "own reputation for good judgment that was at stake. As President of Harvard University his word was representative of American scientific scholarship. To guess wrong could mean incalculable damage to the nation's respect for its men of learning." [28] If incalculable damage of the sort feared by Compton were the con-

sequence of every major technological blunder that had scientific endorsement, the procession of scientists to Washington would have
ceased long ago. But only a scientist, smitten by the xenophobia and
chauvinism of science, could harbor the anxiety that if Conant had
erred in good faith, his professional community would suffer harsh
consequences.*

The curriculum reform movement referred to above is one manifestation of the scientists' concern for lay appreciation of science. Another
manifestation is the continuing distress that scientists feel about the
quality of popular writings on science. Politicians long ago learned to
expect the vagaries of the press and more or less to take them in stride,
but scientists, with the reverence they feel for their work, are violently
distressed by the frequent incapacity of lay writers to convey accurately
the complexities and subtleties of science. Whereas politicians, with
occasional exceptions, suffer in silence while hoping to fare better in
the next round with the public prints, scientists who have been bruised
by the press often hasten to warn their professional brethren of the
perils—and real they are—of lay scientific reporting. In 1940, for
example, press accounts ludicrously mangled and distorted the results
of a study by George Gaylord Simpson on ancient animal findings in
Montana. Simpson, now of the Harvard Museum, but then of the American Museum of Natural History, had taken pains to point out in a
museum press release that the faunas he described were not in a direct
line of modern primates or man. As the press clippings came in, he
found headlines reading, "Man 'Traced' Back 70,000,000 Years";
"Study of Mammals Brings About New Evolution Theory"; "Western
U.S. Now Held to be Man's Birthplace." Simpson acknowledged the
humorous aspects of the episode, but grimly concluded, "It is fairly
typical of what still happens to scientific news and it has a moral, in
fact several of them, that will be obvious to the reader." [29] What the

* With apparent pride, Compton relates the following conversation with General
Leslie R. Groves, head of the wartime atom bomb project: Groves declared that
"you scientists don't have any discipline. . . . You don't know how to take orders
and give orders." To which Compton states he replied, "You are right, General.
We don't know how to take orders and give orders. But a scientist, if he is a
responsible man, has a different kind of discipline. It is not possible for anyone
to tell a scientist what he must do, for his proper course of action is determined
by the facts as he finds them for himself. Then he needs a different kind of
discipline. He needs to be able to make himself do what he sees should be done
without anyone telling him to do it." (Arthur H. Compton, *Atomic Quest; A
Personal Narrative,* New York: Oxford University Press, 1956, p. 113.)

moral was, he did not say. Twenty-six years later, another scientist, Joshua Lederberg, a Nobel laureate, expressed almost precisely the same distress in connection with a wondrously garbled series of misunderstandings that led to a United Press story beginning: "A Sydney University physicist said today the United States, Britain and Russia are studying the possibility of using a nuclear rocket to prevent an asteroid from smashing into the earth." Lederberg reported that after reading the news accounts he wrote to the Sydney physicist to obtain an explanation. The physicist replied that a reporter, in querying him about a recently published article (not of his own writing) on the asteroid Icarus, had asked whether it would be possible to affect its orbit. "My laughing reply," the physicist explained to Lederberg, "was that if, in the dim, distant future, this were deemed necessary, presumably by that time something might be attempted." Subsequently, there appeared the wire service tale conjuring up studies of an Anglo-Soviet-American effort to employ a nuclear rocket against Icarus. A silly, though unforgivable, bit of sensational journalism—one that, in fact, was rejected as nonsense by all but a few newspapers. But from the episode Lederberg drew the following somber conclusion: "Unfortunately, such sensationalism goes on all too often in science news-reporting and erodes the confidence of serious-minded scientists in speaking to the press. This hardly helps the public form an accurate picture of scientific enterprise or of its findings; it only degrades journalism." [30]

Laymen, with the mixture of resentment and reverence that they often feel toward "experts," frequently dote upon cases of scientific amateurs triumphing over the professionals. But that science, with its complexities and requirements for arduous training, its ascent from witchcraft and its indisputable success, should be chilly or even offensive toward the intrusions of outsiders—that is not at all surprising. All skilled groups —whether they be carpenters, fishing guides, or surgeons—have, and often for sound reasons, doubts about the value of amateur opinion.

If all lay contentions about things scientific were to be given careful consideration by the established scientific community, it is quite probable that some worthy work would be discovered. But it is also certain that the literature would be cluttered with nonsense and many productive scientists would be diverted from promising lines of inquiry to knocking down crank theories and witch-doctor remedies. Quite sensibly, the institutions of science do not function on a one-man one-vote basis. Or, as Luis Alvarez puts it, "There is no democracy in physics. We can't say that some second-rate guy has as much right to opinion as Fermi."

But how do the institutions of science function to advance truth, weed out error, honor the worthy, and reject the crackpots? The answer is that, over a period of three and one half centuries, science has evolved an intricate process of certification, no less labyrinthine than the western process of jurisprudence, at least as well revered by its practitioners, and probably no more often in error. (Science, too, occasionally, hangs the innocent and acquits the guilty, but, unlike the law, enjoys the luxury of reversing its error.)

As is the case with virtually all professions and some trades, one is admitted to the ranks of science only upon the approval of those who have previously been admitted to the ranks. A farmer is anyone who designates himself a farmer and who earns a living farming. The more learned callings, of course, do it differently. A would-be lawyer is not a lawyer until other lawyers certify his ability. Similarly, with exceptions so rare that they make the newspapers, a would-be scientist is not certified as a scientist until he obtains a degree under the supervision of previously certified scientists. In the days of basement science, this stamp of approval had considerably less significance than it has today. The homemade scientist could engage in research and, if he was concerned about recognition, take his chances on getting it. Today, however, virtually all research is extremely expensive business; even the theoretician who once needed no more than pencil and paper is being drawn to the costly computer. To obtain access to the tools of science, it is almost invariably necessary to be in the employ of a research institution, and almost invariably, the entree to such employ is an advanced degree.* Thus, the degree is necessary—but it is by no means

---

* There is a certain amount of convenience, if not wisdom, underlying this system. The public and private agencies that provide financial support for research now receive so many applications for grants that, whether or not they publicly acknowledge it, they often, in fact, rate the applicant according to the institution that employs him. Thus, an application from an unknown young scientist at Harvard Medical School is likely to receive a more sympathetic reading than an application from a freelance basement scientist, on the assumption that if Harvard hired him, he must have considerable merit. Furthermore, putting research money into universities serves the added purpose of assisting education in the sciences, since students are usually employed as research assistants. Thus,

sufficient for pursuing a scientific career. Possession of the degree brings with it membership in the community, but it does not bring with it the resources to conduct research. The costliness of contemporary science dictates that another barrier must be passed before a given research project is financially underwritten. The scientist who seeks funds for research will almost invariably have to prepare a proposal for financial support. Despite the lay impression that science runs on the blank checks, someone—in his own institution if he is seeking local funds, or in Washington if public funds are sought—will scrutinize the proposal and approve or reject it. Who does this? Generally, his peers, other scientists who are familiar with his field of research. If they deem it worthy, and money is available, he is past the second barrier, and is in a position to proceed with his research. Now, however, he encounters another barrier that science has erected for the purpose of blocking error and advancing truth. To advance knowledge and obtain credit for that advance, the scientist must make his findings known. But the folkways of the scientific community dictate that the initial exposure of his findings must be in one of the certified forums of science, principally an established journal or a conference of peers. Initial publication in the lay press is considered incompatible with the dignity of science and the maintenance of scientific quality. The editor of *Physical Review Letters,* the leading journal in the physical sciences, has warned publicity-seeking physicists: "As a matter of courtesy to fellow physicists, it is customary for authors to see to it that releases to the public do not occur before the article appears in the scientific journal. Scientific discoveries are not the proper subject for newspaper scoops, and all media of mass communications should have equal opportunity for simultaneous access to the information. In the future we may reject papers whose main contents have been published previously in the daily press." [31] If the object of publication is simply to convey information, why, then, would the lay press not serve? The answer is that lay publications do not incorporate still another barrier devised by the

common sense and bureaucratic simplicity underlie the system, but injustices do occur, as, for example, in the case of William P. Fox, a New York City Police lieutenant, who acquired a Ph.D. in chemistry in his spare time. Fox, whose teachers at Columbia describe him as a very good chemist, chooses to remain in the Police Department, and do his research in his spare time at home. Federal granting agencies consider this quite eccentric (as it is, in terms of the usual behavior of those who possess Ph.D.'s in chemistry) and have declined to support his requests for research support. See *Science,* October 30, 1964, pp. 621–23.

scientific community for separating truth from nonsense: the referee system. When a paper arrives at a scientific journal, the conventional process calls for it to be examined by scientists, on or off the staff, who are knowledgeable in its subject matter, for the purpose of obtaining an assessment as to whether it is original, comprehensible, methodologically sound, and an actual contribution to the advancement of knowledge. In a sense, the referees function as a grand jury. Since they are generally busy with their own scientific work and referee without remuneration, it is not expected that they will subject the prospective publication to verification in the laboratory. In effect, they are merely certifying whether the paper appears to merit exposure to the scientific community. In 1966, for example, *Science* received three thousand articles and fifteen hundred technical comments, letters, book reviews, and meeting reports. With few exceptions, all that was published from this mass was done so on the advice of outside reviewers—generally at least two in the case of each contribution. This process involved the services of some four thousand reviewers.[32] The final barrier is reached when the paper is published and is available for scrutiny by all concerned parties in the scientific community. The prescribed form of publication requires the author to describe his experiments and present the data that he obtained from them—which amounts to a standing invitation for skeptics to prove him wrong. If the work is not refuted, it passes into the accepted body of scientific knowledge and becomes a building block from which other research may proceed.

Such, in brief outline, is the rigorous methodology of science, an elaborate system of checks and screens that is so distant from any counterpart in the nonscientific affairs of mankind that the scientists' reverence for their professional ways is readily understandable.

Does the methodology invariably work? Of course not. And there is ample testimony from the ranks of distinguished scientists to prove that. Max Planck, for example, on the basis of his own experience, sadly concluded: "A new scientific truth does not triumph by convincing its opponents and making them see the light, but rather because its opponents eventually die, and a new generation grows up that is familiar with it." [33] A paper submitted to the British Association for the Advancement of Science by Lord Rayleigh, but from which his name had been omitted or accidentally left off, was rejected, his son and biographer wrote, "as the work of one of those curious persons called paradoxers." When the authorship was discovered it was accepted, causing his son to observe, "It would seem that even in the late 19th century, and in spite of all that had been written by the apostles of free discussion, authority could prevail when argument had failed." [34] The system has never been immune from honest error, deceit, or partisanship. But it is probably more immune from honest error, deceit, or partisanship than any other methodology for establishing truth in this puzzling world. And this perhaps explains the loyalty that scientists so fervently feel toward their ways of doing business. They are so fervent, in fact, that even when the system fails them, they see merit in the failure. Take the case of Polanyi's theory of adsorption of gases, which was rejected by Einstein, among others, shortly after the publication in 1914 ("Einstein and Haber decided that I had displayed a total disregard for the scientifically established structure of matter"). Thirty years later, the theory began to come into acceptance, and Polanyi, widely recognized and honored for other scientific achievements, pondered what lesson might be drawn from this failure of science's elaborate structure of validation. "Could this miscarriage of the scientific method have been avoided?" he asked. "I do not think so. There must be at all times a predominantly accepted scientific view of the nature of things, in the light of which research is jointly conducted by members of the community of scientists. A strong presumption that any evidence which contradicts this view is invalid must prevail. Such evidence has to be disregarded, even if it cannot be accounted for, in the hope that it will eventually turn out to be false or irrelevant. . . . I am not arguing against the present balance between the powers of orthodoxy and the rights of dissent in science. I merely insist on acknowledgment of the fact that the scientific method is, and must be, disciplined by an orthodoxy which can permit only a limited degree of dissent, and that such dissent is fraught with grave risks to the dissenter." [35]

NOTES

1. "Control of the Activity of *Escherichia coli* Carbamoyl Phosphate Synthetase by Antagonistic Allosteric Effectors," *Science*, December 23, 1966, p. 1572.
2. Quoted in James B. Conant, *Modern Science and Modern Man*, The Bampton Lectures, Columbia University, 1952 (Garden City, N.Y.: Masterworks Program, 1952, by arrangement with Columbia University Press), p. 17.
3. George B. Kistiakowsky, "On Federal Support of Basic Research," *Basic*

Research and National Goals, report to the Committee on Science and Astronautics, House of Representatives, by the National Academy of Sciences, 1965, p. 171.

4. Henrik W. Bode, "Reflections on the Relation between Science and Technology," *Basic Research and National Goals,* pp. 46–47.
5. *AEC Authorizing Legislation,* Joint Committee on Atomic Energy, 1964, p. 137.
6. Vannevar Bush, *Science, The Endless Frontier* (1945; reprinted by National Science Foundation, 1960), p. 10.
7. Harvey Brooks, "Future Needs for the Support of Basic Research," *Basic Research and National Goals,* p. 87.
8. *First Report on Project Hindsight,* Department of Defense, June 30, 1966.
9. J. Herbert Hollomon, "Science and Engineering—Fact and Fiction," William C. Ferguson Memorial Lecture, Washington University, November 2, 1966.
10. Quoted in Richard Carter, *Breakthrough, The Saga of Jonas Salk* (New York: Trident Press, Inc., 1966), p. 299.
11. Edward Teller, "The Role of Applied Science," *Basic Research and National Goals,* p. 260.
12. Alvin M. Weinberg, "But Is the Teacher Also a Citizen?", *Science,* August 6, 1965, pp. 601–6.
13. Philip H. Abelson, "Pressure on Basic Research," *Science,* July 1, 1966, p. 11.
14. Harvey Brooks, *op. cit.,* pp. 84–85.
15. "Physics and Man's Understanding," address, September 17, 1965.
16. Alvin M. Weinberg, "Impact of Large-Scale Science on the United States," *Science,* July 21, 1961, p. 164.
17. *The Autobiography of Charles Darwin,* Nora Barlow, ed. (London: William Collins Sons & Co., Ltd., 1958), p. 115.
18. George W. Beadle, "An Introduction to Science," *Listen to Leaders in Science,* Albert Love and James Saxon Childers, eds. (Atlanta, Tupper and Love, Inc., 1965), p. 5.
19. Glenn T. Seaborg, "Chemistry," *Listen to Leaders in Science,* p. 31.
20. *Report of the International Clearinghouse on Science and Mathematics Curricular Development,* Commission on Science Education of the American Association for the Advancement of Science and the Science Teaching Center, University of Maryland, 1966.
21. Weinberg, "But Is the Teacher Also a Citizen?" p. 604.
22. Raymond H. Kahn, letter to *Ward's Bulletin,* December 1966.
23. Ralph E. Juergens, "Minds in Chaos: a Recital of the Velikovsky Story," *American Behavioral Scientist,* Vol. VII, No. 1 (September 1963), p. 7.
24. *Ibid.*
25. Bernard Barber, "Resistance by Scientists to Scientific Discovery," *Science,* September 1, 1961, pp. 596–602.
26. McDonald and Howell, exchange of letters, *Saturday Review,* May 7, 1966, p. 43.
27. "Letters to the Editor," *New York Times Book Review,* June 18, 1967.
28. Arthur H. Compton, *Atomic Quest, a Personal Narrative* (New York: Oxford University Press, 1956), p. 104.
29. "The Case of a Scientific News Story," *Science,* August 16, 1940, p. 148.
30. Joshua Lederberg, "World's End Postponed," column in the *Washington Post,* "Science and Man," September 4, 1966.

31. S. A. Goudsmit, *Physical Review Letters,* 4:109 (1960).
32. Philip H. Abelson, *"Science* and the Scientific Community," *Science,* September 23, 1966, p. 1473.
33. Quoted in Bernard Barber, *op. cit.*
34. *Ibid.*
35. Michael Polanyi, "The Potential Theory of Adsorption," *Science,* September 13, 1963, pp. 1010–13.

BOOK
TWO

III

When Science
Was an Orphan

. . . hardly anyone in the United States devotes himself to the essentially theoretical and abstract portion of human knowledge.

TOCQUEVILLE, in *Democracy of America*

Unemployment among scientifically and technically trained men has been, and is, acute. The plight of the technically trained young men, who have received an expensive education and who are essential to the future life of the country, is pathetic.

KARL T. COMPTON, President, Massachusetts Institute of Technology, and ALFRED D. FLINN, Director, Engineering Foundation, in *Recovery Program of Science Progress*, 1933

We have depended a great deal in the past for our basic research and some of our applied research on the overtime of tired professors and instructors who carry a heavy load of teaching. . . .

HARLOW SHAPLEY, Director of Harvard Observatories, before the Senate Committee on Military Affairs, 1945

It has been argued that since World War II, science has become the pawn of government, and then there are those who contend that government has actually become the pawn of science. But there is no question that over the past twenty-five years, the two parties have progressed to an energetic intimacy that represents a virtually total departure from the practices and traditions of prewar days. Against the contemporary background, it may be difficult to realize that prior to 1940, not only was there a mutual aloofness between the federal government and the most influential and creative segments of the scientific community, but there

51

were strong feelings on both sides that the separation was desirable. In maintaining a distance from science, the federal government was not out of harmony with the anti-intellectual strain that pervaded American life. And in being distant from government, the scientific community was not only bowing to the fact that it was not wanted, but was also pridefully responding to rebuffs and crude treatment that it considered inimical to its integrity. Science was the first to recognize that it needed government's support and that government needed its skills, but the recognition came slowly, and on the eve of World War II, mutual aloofness remained the dominant theme. To comprehend the internal and external politics of contemporary science, it is first necessary to trace the prewar setting, and then to examine the conditions and processes that dissolved a pattern that had endured for a century and a half.

Prior to World War II, while basic science existed more or less as an orphan on the American landscape, applied science and technology were accorded an esteemed place and rather generous support. In a society keyed to economic progress, honor and financial reward were bestowed upon the tinkerer, the gadgeteer, the Yankee engineer, those, in short, who could translate knowledge into utility. Edison, the inventor, was more of a folk hero than Compton, the physicist. Only a dozen years before the outbreak of World War II, Herbert Hoover estimated that from all sources, the nation was spending a total of $200 million a year in the applications of science, but only $10 million in pure research; that 30,000 scientists and engineers were employed in applying science, but less than 4000 were engaged in pure research, "most of them dividing their time between it and teaching." Hoover added, "For all the support of pure science research we have depended upon three sources—that the rest of the world would bear this burden of fundamental discovery for us, that universities would carry it as a byproduct of education, and that our men of great benevolence would occasionally endow a Smithsonian or a Carnegie Institution or a Rockefeller Institute." [1] Hoover's lament ignored the role of industry as a source of support for basic research, but his assessment was generally valid. From the earliest days of the Republic, not only did technology thrive in the open marketplace, but also there were few reservations about the desirability or the political legitimacy of government promoting and employing technology for specific public purposes, such as exploring and surveying the westward lands, assisting navigation, developing standards for weights and measures, and carrying out quarantine and other public health measures. Technology was clearly too useful and potent to be ignored by government, even in a nation emotionally and constitutionally dedicated to circumscribing the influence and activities of the central power. Thus technology, because of its readily apparent utility, was early and easily welcomed into our national life. But a far different reception awaited basic research, both in the nation at large and in the federal government.

The attitudes that motivated this reception were not elicited by an aversion or lack of interest exclusively confined to science. Rather they were the product of far-ranging public attitudes that thwarted even the political intellectuals and scientific scholars who dominated the first years of the nation's history, Jefferson, Franklin, Madison, and the Adamses. When Washington, for example, proposed the establishment of a National University, even the anti-centralist Jefferson lent his support, and suggested that the university acquire the faculty of the financially troubled University of Geneva. But Congress saw no need for the federal government to venture into education. Toward the middle of the century, it took the Congress nearly a decade to accept the bequest of James Smithson, the British chemist who left his fortune to the United States for the establishment of an institution for "the increase and diffusion of knowledge among men." In the 1840's, when Congress was still gagging on the Smithson bequest, fundamental research in Great Britain and Europe was aglow with the works of Dalton, Faraday, Gauss, and Helmholtz, and German universities were embarking on their great period of scientific creativity. On both sides of the Atlantic, science often was a gentleman's profession, or, in the case of medical research, derived support through medical practices. In the Old World, however, basic science could also draw economic sustenance from princely and commercial patronage, or through government-supported universities and research institutes. But in the New World, there were no princes, industry was yet to recognize the profitability of science, and government was concerned with utility, not fundamental knowledge.

There is, however, a danger of overdrawing the picture, to the effect of portraying the Old World as single-mindedly dedicated to the pursuit of fundamental research, while any inquiry not visibly related to utilitarian purposes was thwarted in the United States. Such a picture is nonsensical. The American Philosophical Society, for example, came into being in 1743; the American Academy of Arts and Sciences, in 1780; *Medical Repository,* generally considered to be the first scientific journal in the United States, was founded in 1798. The Smithsonian Institution, after its extremely painful birth in 1846, promptly elected Joseph Henry, the nation's most distinguished physicist, as its chief executive, and the institution's Board of Regents declared, "The increase of knowledge by original research shall form an essential function" of the Smithsonian Institution.[2] Meanwhile, one of the great luminaries of basic research during that period was complaining, "Our ministry still clings to the fiction that it will be involved in war in the East, and declines to make any outlay of money." [3] The words were written by Helmholtz during the period 1855–58, when he was professor of anatomy and physiology at Bonn.

Nevertheless, once some reasonable sense of perspective is achieved, the fact is that, while the Old World was receptive and encouraging to the conduct of fundamental scientific inquiry—and, in one way or another, allocated relatively generous economic resources to such activity —the New World rarely offered encouragement, and, in most instances, merely tolerated the presence of pure science. Whether during this period of our national development the emphasis on utility was, in fact, a wise economic choice, is a separate and difficult question, for, as noted previously, the relationship of science to technology is far from clearly understood; going one step further, the relationship of technology to economic development is befogged by such variables as capital investment, cultural attitudes, natural resources, the political setting, and educational levels. But, for purposes of understanding the contemporary politics of science, it is necessary to recognize this key economic fact of the pure scientists' legacy from prewar days, namely, that relative to the situation in Great Britain and Western Europe, pure science in the United States had to scratch hard for a living.

Tocqueville concluded that the Americans excelled at deriving utility from basic knowledge imported from Europe. But, he noted, "It must be acknowledged that in few of the civilized nations of our time have the higher sciences made less progress than in the United States. . . ." [4] His observation denied proper respect to Joseph Henry, Benjamin Franklin, and other American scientists for their enterprise in fundamental research. But in essence, the observation was correct. The support and esteem that government and private sources provided for utilitarian research tended to draw the technically talented in that direction, while those with the taste or capacity for fundamental research found scarcely any support.*

A. Hunter Dupree points out in his pioneering study, *Science in the Federal Government,* that one of the earliest instances of federal support for research in a nongovernment institution was a grant in 1830 for the Franklin Institute to investigate the causes of boiler explosions.[5] This was for applied research, inquiry directed toward a specific utilitarian objective. But Congress was opposed to the establishment of a government-supported astronomical observatory, and it regularly foiled the efforts of those scientists who sought the creation of permanent federal bureaus to support scientific inquiry. While private funds for research were in minute and uncertain supply until the creation of major philanthropic foundations at the beginning of the twentieth century, scientists in the employ of the government's technical bureaus became adept at surreptitiously diverting to basic research money that was earmarked for utilitarian programs. In 1841, for example, Congress still balked at the Navy's desire to construct an astronomical observatory. But the head of the Navy's Depot of Charts and Instruments prevailed upon a House committee to authorize $25,000 for a building for studies in hydrography, astronomy, magnetism, and meteorology—all depicted as of practical value to the Navy. The resulting structure was called the Depot of Charts and Instruments, but the architecture met the requirements of astronomical observation, causing John Quincy Adams, long a proponent of a government observatory, to state that he was "delighted that an astronomical observatory—not perhaps so great as

* The popular impression of a vigorously thriving Yankee inventiveness from the very beginnings of American history actually deserves a good deal of qualification. As Reynold M. Wik points out in an essay, "Science and American Agriculture" (*Science and Society in the United States,* edited by David D. Van Tassell and Michael G. Hall, Homewood, Ill. [Dorsey Press, 1966]), "Since the period from 1607 to 1775 represents almost half of the total span of United States history . . . we would be entitled to expect the colonists to have produced some remarkable machines prior to the Revolutionary War. Unfortunately, however, evidence now suggests that the colonial people failed to invent *any*—successful machines of any importance; at least historians fail to mention them. . . . Benjamin Franklin is often called America's first inventor with his improved stove, lightning rods, electrical devices, and his rocking chair with a built-in fan to keep off the flies. While these innovations were useful, none of them could properly be classed as a mechanical machine, as an apparatus of interrelated parts with separate functions and capable of performing some kind of work. . . . Indeed, the most striking feature of colonial technology is its sterility, not its inventiveness."

it should have been—had been smuggled into the number of the institutions of the country, under the mask of a small depot for charts . . . when a short time before a provision had been inserted in a bill passed, that no appropriation should be applied to an astronomical observatory." [6] It is interesting to note that at virtually the same time that the Navy was "smuggling" an observatory past Congress, Czar Nicholas was pressing the construction, without regard to cost, of what was, for the time, the best-equipped astronomical facility in the world, the Pulkova Observatory.[7]

Thus, Congress, in harmony with the general temper of the American people, early demonstrated its dedication to a principle that would not finally disappear until the turmoil of World War II swept away many well-established patterns: It was frequently willing, sometimes eager, to employ public funds and authority to apply knowledge for public purposes, but it yielded only slowly in its belief that public funds should not be used to support the acquisition of knowledge. In 1885 a debate could rage over how the federally subsidized agricultural research stations should divide their efforts between fundamental research and research applicable to the immediate problems of the farmer. *Science* commented, "There appears to us to be comparatively little danger that the work of American experiment stations will be too rigidly scientific. . . . There is a constant pressure upon a station for immediately useful results, and any station refusing reasonable conformity to it will not enjoy a long life." [8] In 1965, the very same issue was at the heart of the matter when a Senate committee reviewed government-supported research in weather modification, and concluded that basic research had proceeded long enough and it was time to make some rain. The difference was, however, that in 1965, basic research was so accepted in the political process that the Senators' sense of frustration was not reflected in their budgetary decisions.[9]

The distinction between knowledge and utility was to have profound effects on the progress of American science. With government primarily interested in the utility of science, and with industry, the other great potential source of support, similarly concerned with usefulness, basic research in the United States became of necessity a part-time occupation, carried on by amateurs or academics, who could justly complain that their research efforts were impaired by the necessity of earning a living.

Starting about the last quarter of the nineteenth century, these complaints as well as a number of other factors slowly—but very slowly—began to force change upon this situation. First, American students with scientific or medical ambitions were increasingly attracted to study abroad, particularly in Germany, where competition among a score of state-supported universities had created a hothouse atmosphere for scientific and other scholarly pursuits.[10] In the German universities they observed that research was accorded especial prestige, that concentration on it was encouraged with full-time chairs and other financial support. Furthermore, the appreciation of science in Europe had matured to the point where researchers could find support for specialized institutes, such as the Pasteur Institute, founded in 1888, and the Koch Institute, in Berlin, founded in 1891. In the last decade of the nineteenth century, some three hundred German scientists were primarily engaged in biological and medical research.[11] In the United States, the number who could concentrate on research was minuscule, and there was a virtual absence of anything resembling research chairs, except for a few in astronomy and preclinical medicine. By one account, there were in the 1890's only twenty-six graduate fellowships available for all the university departments in the country.[12] The contrast with Europe was a shocking one, and inevitably produced pressures for bringing the European, but particularly the German, research orientation into American higher education. The pressures were further increased by the great disparity in quality between American- and European-trained medical practitioners. The University of Michigan, in 1851, was the first to put its medical professors on a salaried basis, but the emulation of European methods was slow to take hold, and it was not until the founding of Johns Hopkins University in 1876, and particularly the Hopkins Medical School in 1893, that research, as a principal occupation, began to obtain significant recognition in American universities.[13]

As this change took place, it was given impetus by a number of other developments. Seeking some socially desirable objectives for the wealth they had captured, Rockefeller and Carnegie fixed upon the intellectual realm—with special emphasis on science—and poured forth amounts that, for the time, were quite extraordinary. The Rockefeller Board had a mere $20,000 to dispense for medical research grants in 1901, but the next year it had $1 million. Soon after, it evolved into the Rockefeller Institute for Medical Research, for which Rockefeller in 1907 provided an endowment of $2,620,000. By 1920, this was increased to $23 million, and by 1928, to $65 million.[14] Meanwhile, Carnegie established the Carnegie Institution of Washington, provided it with $10 million in 1902, an additional $2 million in 1907, and another $10 million in 1911.[15]

Simultaneously, another channel of support began to develop. Industry slowly realized that it could be profitable to free basic researchers from pressures to turn a profit, that puttering that seemed aimless to the layman might, in fact, lead to understandings or developments of great commercial value. General Electric and American Telephone and Telegraph pioneered in the establishment in this country of major industrial research organizations, but it must be added that they pioneered slowly, and that the rest of American industry moved even more slowly. Even the industrialization of human slaughter in World War I failed to establish any significant reliance of government upon science. James B. Conant, for example, recalled that when the American Chemical Society offered its services to the Secretary of War upon this nation's entry into that conflict, the Secretary replied "that it was unnecessary as he had looked into the matter and found the War Department already had a chemist." [16] Immediately following the war, General P. C. March, the War Department's Chief of Staff, wrote, "Nothing in this war has changed the fact that it is now, as always heretofore, the Infantry with rifle and bayonet that, in the final analysis, must bear the brunt of the assault and carry it on to victory." [17]

Nevertheless, no matter how strong or influential the aversion to governmental entanglement with science, the functioning of government required the existence of technical bureaus. And once they were in business, no matter how diligent Congress was to keep them devoted to visibly useful work, the scientists' yen for basic research was irrepressible—even if it meant bootlegging basic research into supposedly utilitarian programs. Occasionally there were blowups, as in the late nineteenth century, when Department of Agriculture botanists had their own ideas about what lines of research would be most beneficial. As Dupree points out, "the botanists tended to concentrate on what the department could do for their science instead of what it could do for the farmers." [18] But whether or not the politicians liked it, science was inevitably seeping into the federal bureaucracy. In 1915, the nation's inadequacy in aviation research and production—World War I found the United States with only 23 aircraft compared with 1400 in France and 1000 in Germany [19]—led to the establishment of the National Advisory Committee for Aeronautics. Administratively, NACA was a unique hybrid, composed of government and nongovernment members, authorized to dispense government funds for research in federal and nonfederal institutions. The amounts were small, but the precedent was large. Similarly, by the late 1930's, the popularity of medical research went beyond the resources of the universities and the private philanthropic foundations, and Congress authorized the creation of the National Cancer Institute, which between 1938 and 1940, dispensed $220,000 in grants.[20]

As welcome as this blossoming of support for basic research may have been, the fact is that right up to the beginning of World War II, financial malnutrition remained a serious malady throughout the American scientific community. And the affliction was reflected in the productivity of American science. Clearly, the productivity had been improving, particularly in the medical sciences, which, of all the sciences, were, and still remain, the favorites of private philanthropy. But it was not until the vast infusion of federal funds in the postwar period that American science dramatically began to outstrip its European counterparts. Prior to World War II, the Germans' lead in medical research was overcome not so much by a surge of American effort as by the shattering of German research institutions in the first World War, the ensuing political chaos, and the flight of German Jewish scientists. Attempts to assess the significance and national origins of scientific discoveries encounter so many uncertainties that they must be viewed merely as impressions of trends, rather than precise evaluations. But according to one compilation, between 1880 and 1889, out of a total of 147 significant medical discoveries, American scientists produced 18, while German scientists produced 74. Between 1900 and 1909, the respective national figures were 28 and 61, out of a total of 148. In the period that included World War I, 1910–19, the pattern was reversed, with the United States producing 40 and Germany producing only 20, out of a total of 99. In 1920–26, out of a total of 44 discoveries (in part, probably, the totals were dropping because progress was bringing research to more difficult problems) the United States dominated the field, with 27 discoveries, compared to 3 for the once-dominant Germans.[21] In the distribution of Nobel prizes, the most carefully bestowed and prestigious awards in basic research, there is a striking illumination of the pre- and postwar position of American science relative to the once-dominant

European science. The prizes, in physics, chemistry, and medicine and physiology, were first awarded in 1901. By 1939, Americans had received only 15 of the 128 awards; between 1943 and 1956, they received 34 out of 67.[22] Just prior to World War II, Americans seeking an advanced scientific education still went abroad in large numbers. And, in fact, that most conspicuous wartime triumph of American technology, the atomic bomb, was a product of almost purely European-produced fundamental principles. The scientists whose efforts were most crucial to the success of the bomb project were in large part foreign-born and -trained or were Americans who learned much of their craft in European laboratories. If the task of building the bomb for use in World War II had been confined to an all-American expertise, it most probably would have been hopeless.

The extent to which science should be supported involves multifarious complexities, most of which did not present themselves for serious contemplation until the early 1960's, when research suddenly emerged as a large and readily visible slice of the federal budget. It was then that "research on research" became a thriving occupation, dealing with a variety of perplexing issues, including whether science might incur financial gout, or whether a society might overinvest in science, to the detriment of society and science.* Between the world wars, however, though support of science was generally in a continuous increase, there was no occasion for examining such wholesome problems. By any enlightened standard, it would have to be concluded that American science in that period was grossly undersupported. Talented students were financially unable to obtain scientific training; talented scientists were unable to find employment in science. Those who were employed as scientists were often severely handicapped by inadequate funds for supplies and equipment. Promising lines of scientific inquiry were neglected for lack of funds.

Against this background, it must be noted that the federal government still remained aloof from, and, in fact, often hostile to basic science. But for the purposes of our understanding of the politics of

---

* One measure of the dynamic state of the "research on research" industry is an inventory compiled by NSF, *Current Projects on Economic and Social Implications of Science and Technology,* 1965 (NSF 66-21). Covering 187 pages, it divides the subject matter into 14 subdivisions, and lists several hundred individual research projects under way at universities throughout the country.

science, it is even more significant to note that science had no great affection for government, nor for any other source of support that threatened the possibility of intellectual domination through financial control. Whether or not science was as vulnerable to contamination or control as its leaders made it out to be is another separate and difficult question. But in any case, through a combination of sorry experience and fervent imagination, the profession of science had acquired a mystique of intellectual fragility and a sensitivity to what it considered to be threats, or even the possibility of threats, to its integrity and virtue. At the turn of the century, the financially weak Woods Hole Marine Biological Laboratory could proudly assert that it had held to its concept of the practice and teaching of marine biology at the price of declining assistance from the newly founded Carnegie Corporation and a group centered at the University of Chicago.[23] Nearly half a century later, Frank Jewett, president of the National Academy of Sciences and one of the most chauvinistic inhabitants of the scientific community, proudly related the incident before a congressional committee that was examining proposed arrangements for government support of research. The Woods Hole Laboratory, Jewett explained to his lay audience, "is the greatest institution in the world in the field of the development of fundamental biologists. . . . For the first fifteen or twenty years of its life it was a starving institution. The lowly paid college professors camped in tents and lived in huts and put up a little bit of money. They needed money. They needed it like everything." But when Carnegie and the Chicago group offered them money, they were so fearful of the "conditions which were imposed on the money that they turned those offers down—when they were starving for money. That," concluded Jewett, "was the attitude of those people of science at that time with regard to anything which dominated." [24]

Self-righteous purity? Priggishness? Realism? Probably some of each. Yet the folklore of scientific independence rested firmly on the painful history of intellectual persecutions, from Socrates to Galileo, from the derision of Darwinism to the book burnings of the Nazis. Better no dependence upon government than a dependence that would allow crude interference, let alone control. This was the dominant ideology of American science in relation to the outside world, in sharp contrast to the European experience, where government patronage was quite early accepted as a means through which support could easily be reconciled with the requirements of scientific independence. The Venetian

Republic was the source of support for the University of Padua, where Galileo was appointed to the chair of mathematics in 1592.[25] The Cambridge of Newton received Royal funds. In 1675, Charles II established the Royal Observatory at Greenwich. In modern times, European nations evolved a variety of mechanisms for providing government funds for nongovernment science and technology, ranging from government-subsidized industrial research associations, first started in 1915 upon recommendation of the Advisory Council for Scientific and Industrial Research in Britain, to the post-World War I German Research Council, established in an attempt to restore Germany's scientific leadership.[26] But these foreign experiences had little direct effect on the mutual aloofness that had long characterized the relationship between the federal government and American practitioners of basic science. Applied science was well established within the province of government; basic science, when it took place with government support, was often a bootlegged affair. Its primary setting was the universities, which, with the exception of the land-grant colleges, operated without federal assistance. And in these institutions, first established under the Morrill Act in 1862, research was concentrated on agricultural matters, and almost invariably the emphasis was on applied research. At the turn of the century, the division of applied science in government and basic research in the universities was, in effect, ratified by a Committee of Organization of Government Scientific Work, appointed by Theodore Roosevelt. In 1903, it concluded that science "on the part of the Government should be limited nearly to utilitarian purposes evidently for the general welfare." Five years later, a committee of the National Academy of Sciences, appointed at the request of the Congress, arrived at similar conclusions.[27]

Beginning around the second decade of this century, the folklore of scientific independence, admirable and inspiring as it might be, was rapidly being eroded by raw economics. Nature no longer had much to reveal to the dilettante or the barehanded investigator. Science had evolved into an institution, staffed by well-trained practitioners, employing expensive instruments. One measure of the sweep of this change is to be seen in an analysis of membership of the Royal Society, perhaps the most selective and prestigious of scientific societies. In 1881, its membership included 54 scientists who could be characterized as "distinguished laymen." By 1914, their number had declined to 38; by

1953, it was 8. In the same years, the clerical members declined from 14, to 4, to 0. Meanwhile, membership from academic institutions rose from 134, to 289, to 348.[28] Research was a full-time expensive enterprise, but how was it to be supported in a nation where private sources were inadequate and government and science were indifferent to each other?

By the eve of the depression the financial condition of American basic research was becoming critical. In 1925, Secretary of Commerce Hoover, whose role as a friend of science has gone largely overlooked, warned: "While we have in recent years developed our industrial research upon a scale hitherto unparalleled in history, we have by no means kept pace in the development of research in pure science." Hoover added that the lure of well-supported applied science was drawing talented men away from pure research, and that in the long run both types of research would suffer from the neglect of basic studies." [29] His remedy for the situation, formulated in conjunction with the National Academy of Sciences, was a proposal for an industry-supported $20-million fund to assist basic research in the universities. Both his appraisal and remedy apparently were in harmony with the feelings of major segments of the scientific community. *Science,* for example, commented that scientists, "Too often with the comfort of their families at heart . . . reluctantly accept well-paid industrial positions instead of poorly paid academic posts. . . . In short, able investigators should be given some . . . comfort in life, freedom of action and opportunity for constructive thought that industrial and administrative officers in this country, certainly of no larger caliber, habitually enjoy." [30] By 1930, however, less than half a million dollars had been subscribed to the fund, and a few years later the proposal was abandoned because of the inadequate response.

Financial desperation now tended to supersede folklore. Having been rebuffed by industry, the leaders of basic research turned to the federal government, and periodically through the depression, petitioned for the rescue of their profession. In 1933, before the New Deal turned to spending as an economic palliative, the Roosevelt administration sought to exceed its predecessor in cutting federal expenditures, including those of the traditionally starved government technical agencies. Meanwhile, the depression had severely eroded the endowments of universities and philanthropic foundations. Typical was the case of the California Institute of Technology, where a $4.2-million endowment, whose income was earmarked for the development of the physics department, was reduced to some $250,000 by the economic crisis.[31] Industry, seeking to stay afloat, widely concluded that it could not afford the luxury of scientific studies with a distant payoff. Science was reduced to an egregiously poor condition. Recommendations for the relief of science were formulated by a Science Advisory Board, created in 1933 under a presidential executive order. MIT President Karl T. Compton, who served as chairman of the board, and Alfred D. Flinn, director of the Engineering Foundation, declared that the plight of American scientists, especially the younger ones, was so "pathetic" that government had a "responsibility to extend its measures for emergency unemployment relief to scientific workers." [32] The Board endorsed a proposal of Compton and Flinn for the government to provide $16 million for the support of scientific and technical research over a six-year period. But the proposal, in that form and in amended versions, drew no more than kind words from the political councils, and the Science Advisory Board expired in 1935, without having brought any relief to American science.* To a large extent the SAB foundered because of strife between the Academy and its supposedly subordinate Research Council, and also because the Academy, which was the statutory scientific adviser to government, was aggrieved by the White House's decision to establish and appoint a new advisory body. The decision sent paroxysms of concern through the Academy traditionalists. For example, upon learning that the SAB membership was to be selected by nonscientist politicians, three distinguished scientists, Robert Millikan, Arthur A. Noyes, and George E. Hale, telegraphed the Research Council president: "We heartily favor establishment of Science Advisory *with Academy membership* [italics supplied]. But are strongly convinced you . . . should be chairman of board . . . [to avoid] possible public inference of new independent science agent of government." [33] But the prospects of the committee were also blighted by the long tradition of aloofness that separated science and government. Government neither understood nor

---

* For detailed accounts of this fascinating early encounter between science and government see Carroll W. Pursell, Jr., "The Anatomy of a Failure: The Science Advisory Board, 1933–1935," *Proceedings of the American Philosophical Society*, Vol. 109, No. 6 (December 1965). Also Lewis E. Auerbach, "Scientists in the New Deal: A Pre-War Episode in Relations Between Science and Government in the United States," *Minerva*, Summer 1965. I am also indebted to Frederick Seitz, president of the National Academy of Sciences, for making available to me records of the SAB.

appreciated science, nor felt any responsibility for the professional well-being of the practitioners of science. And, for its part, science was still well imbued with a strong, though diminishing, aversion to government. Professor E. B. Wilson, of Harvard, directing his remarks to Roosevelt's affront to the Academy, warned that the presidentially appointed Advisory Board "takes away the independence of scientific men or it may take it away and may result in a committee supposedly under scientific auspices being under political necessity of making recommendations politically suitable." [34] One study of the SAB concludes that of its fifteen members "about ten members feared political involvement more than they wished to involve science in society, while only three or four felt the risk of political control was worth taking in almost every case." [35] Furthermore, the SAB members, all natural scientists, saw no reason why the social sciences should be included in their plans for government assistance. Unfortunately for the SAB, however, its amended proposals were referred by Roosevelt to Interior Secretary Harold Ickes, who turned them over for study to a board composed entirely of social scientists. For their part, the social scientists saw no reason why their supercilious colleagues in the so-called "hard" sciences should be singled out for support. Not too many years later, the natural scientists, triumphant because of their achievements in World War II, were to return the compliment when the question of federal support for the social sciences arose.

But in New Deal days, government was not inclined to rush to the assistance of the scientific community. In 1938, the National Resources Committee, chaired by Ickes, noted that "The universities are the centers of 'pure' research." [36] It estimated that the universities spent about $50 million a year on research in 1935–36, and that, of the total, the federal government provided $6 million, "chiefly for research in agriculture." [37] On the eve of World War II, some federal funds began to flow into university research through the Works Progress Administration. But the basic sciences received a trivial portion of this assistance. For example, in its 1938 report, the WPA listed itself as having provided at least some support for 21 projects in mathematics, physics, and astronomy. (Among the scientists associated with these projects were some who had already achieved or who were to go on to great distinction, including two Nobel laureates, Glenn Seaborg and Ernest O. Lawrence.) In chemistry, the agency was associated with 9 projects. But in Educational Sociology there were 70 projects; in Agricultural Technology, 75, and in Motor Car Transportation and Traffic Surveys, 214.[38] Neither WPA nor any other branch of government was disposed to regard basic research as a critically important element of national strength. As far as the scientific community was concerned, its aversion to government had been greatly eroded by a decade of suffering, but the partnership that we know today was to be forged in an unforeseen way, namely, at the outset of World War II, when the leaders of American science, including several alumni of the Science Advisory Board, concluded that whether or not science needed government, government desperately needed science. The unsuccessful supplicant was embarked on saving the rejecting patron, and it can be reasonably argued that the long-time orphan did precisely that, with a display of persuasiveness, political skill, and technical performance that comprise one of the most arresting chapters in the nation's history.

NOTES

1. *Science,* 65:26–29 (1927).
2. Walter Karp, *The Smithsonian Institution* (The Smithsonian Institution in Association with the Editors of *American Heritage* Magazine, 1965), p. 20.
3. Leo Koenigsberger, *Hermann von Helmholtz* (Oxford: The Clarendon Press, 1906), p. 147.
4. *Democracy in America* (Vintage Books; New York: Random House, Inc., 1945), Vol. 2, p. 36.
5. (Cambridge, Mass.: Belknap Press of Harvard University Press), p. 50.
6. *Congressional Globe,* Vol. 15, 1st Session, 29th Congress, p. 738.
7. Sir Harold Spencer Jones, "Science in the USSR: Astronomy and Terrestrial Magnetism," *Nature,* Vol. 156 (September 15, 1945), p. 324.
8. *Science,* 5:21 (1885).
9. *Science,* January 15, 1965, pp. 274–76.
10. Joseph Ben-David, "Scientific Productivity and Academic Organization in Nineteenth Century Medicine," *American Sociological Review,* Vol. XXV, No. 6 (December 1960).
11. Richard H. Shryock, *American Medical Research, Past and Present* (New York: Commonwealth Fund, 1947), p. 89.
12. *Ibid.,* p. 88.
13. John E. Deitrick and Robert C. Berson, *Medical Schools in the United States* (New York: McGraw-Hill Book Company, Inc., 1953), p. 32.
14. Shryock, p. 92.
15. Dupree, *Science in the Federal Government,* p. 297.
16. James B. Conant, *Modern Science and Modern Man,* The Bampton Lectures, Columbia University, 1962 (Garden City, N. Y.: Masterworks Program, 1952, by arrangement with Columbia University Press), p. 18.
17. Quoted in Clarence G. Lasby, "Science and the Military," *Science and So-*

ciety in the United States, David D. Van Tassell and Michael G. Hall, eds. (Homewood, Ill.: Dorsey Press, 1966), p. 261.

18. Dupree, p. 156.
19. Dupree, p. 286.
20. Dupree, p. 366.
21. Joseph Ben-David, *op. cit.,* pp. 828–53.
22. *Basic Research and National Goals,* a report to the Committee On Science and Astronautics, House of Representatives, by the National Academy of Sciences, 1965, p. 165.
23. *The Woods Hole Marine Biological Laboratory* (Chicago: University of Chicago Press, 1944), pp. 48–61.
24. *Hearings on Science Legislation,* Subcommittee of the Senate Committee on Military Affairs, October 1945, p. 492.
25. Ludovico Geymonat, *Galileo Galilei* (New York: McGraw-Hill Paperbacks, 1965), p. 16.
26. Edward McCrensky, *Scientific Manpower in Europe* (New York: Pergamon Press, Inc., 1958), pp. 127–37.
27. Dupree, pp. 294–96.
28. D. S. L. Cardwell, *The Organisation of Science in England* (London: William Heinemann, Limited, 1957), p. 176.
29. *Science,* 63:10 (1926).
30. *Ibid.,* p. 11.
31. Robert A. Millikan, *Autobiography* (New York: Prentice-Hall, Inc., 1950), p. 246.
32. *Report of the Science Advisory Board, 1933–34,* p. 271.
33. Telegram to Isaiah Bowman, Archives of the National Academy of Sciences.
34. Quoted by Lewis E. Auerbach, "Scientists in the New Deal: A Pre-War Episode in Relations Between Science and Government in the United States," *Minerva,* Summer 1965.
35. Auerbach, *op. cit.*
36. *Research—A National Resource: I. Relation of the Federal Government to Research,* National Resources Committee, December 1938.
37. *Ibid.,* p. 189.
38. *Index of Research Projects, Vol. 1, Works Progress Administration, 1938.*

IV

The War-born
Relationship

Military laboratories [prior to World War II] were dominated by officers who made it utterly clear that the scientists and engineers employed in these laboratories were of a lower caste of society. . . . [The] senior officers of military services everywhere did not have a ghost of an idea concerning the effects of science on the evolution of techniques and weapons. . . .

> VANNEVAR BUSH, Director of the World War II
> Office of Scientific Research and Development,
> in *Modern Arms and Free Men*, 1949

Bush did a very great thing by just setting up an organization in which it was possible for a scientist to make military contributions with dignity and effectiveness.

> I. I. RABI, 1965

The impending war in Europe found the American scientific community in a peculiar relationship to its national surroundings. Science was remote from society in general and government in particular. Yet there was a tacit mutuality of sentiment between the anti-isolationist Roosevelt administration and many of the most influential and creative members of the scientific community. Roosevelt was seeking to break the nation away from its isolationist complacency, an objective which aroused widespread and fervent political opposition. At the same time, the scientific community was one of the, if not the most, internationally acquainted segments of American society. In its substantive proceedings, science was necessarily oblivious of international boundaries. In astronomy, geology, zoology, and in virtually every other discipline, the distribution of scientific data did not conform to political divisions. Thus, the practical requirements of the conduct of research gave science a supranationality that distinguished it from most other professions. The professional training and problems confronting an American lawyer, for example, would generally bear little, if any, relationship to those of a French counterpart. But an American physicist and his French counterpart were regularly addressing themselves to the same problems, might well have learned their craft under the same German master, communicated with each other through the same professional journals, visited each other's laboratories, and met at international scientific congresses. The deep roots of internationality in science is illustrated by a letter written during the American Revolution by Benjamin Franklin when he was envoy from the Congress of the United States to the French Government. Empowered to issue letters of marque against British vessels, Franklin specifically instructed American ships not to interfere with the British Captain Cook, who was shortly expected in European waters, en route home from his third great voyage of Pacific explorations. Through Cook's explorations, Franklin told naval commanders bearing American war commissions, "the common enjoyments of Life are multiply'd and augmented, and Science of other kinds increased to the benefit of Mankind in general; this is, therefore, most earnestly to recommend to every one of you, that, in case, the said Ship . . . should happen to fall into your Hands, you should not consider her as an Enemy, nor suffer any Plunder to be made of the Effects contain'd in her, nor obstruct her immediate Return to England . . . but that you would treat the said Captain Cook and his people with all Civility and Kindness, affording them, as common Friends to Mankind, all the assistance in your Power . . ." [1] France, though in alliance with the United States, issued similar instructions to its Navy and privateers, as did the Spanish government. Though the products of science might be utilizable for international strife, the practice of science, and the men of science, partook of a virtually unique supranationality.*

* Arnold W. Frutkin, director of NASA's Office of International Programs, argues in his extremely interesting book, *International Cooperation in Space* (Prentice-Hall, 1965, pp. 10–17), that the internationality of science involves more form than substance, that when the brotherhood of science conflicts with national interests, the latter invariably dominates. Franklin's directive, he notes, was never put to the test, since Cook's vessel—Cook was actually dead when Franklin wrote his order—returned home without being intercepted. In other cases, Frutkin observes, safe conduct for scientists was at times honored and at times ignored, leading him to the conclusion that such incidents "tell us little more than that there were gentlemen and nongentlemen in war. Consideration and generosity have never been the special province of scientists, and prisoner exchanges have been common among all sorts of people." Frutkin correctly adds that during the Civil War, the National Academy of Sciences exacted a loyalty oath of its members, and finally concludes: "The evidence appears to be over-

With the improvement of communication and the burgeoning of scientific effort on both sides of the Atlantic, there existed at the outbreak of World War II a large and truly international scientific community, and perhaps no segment of it, at least as far as basic research was concerned, was quite as international as nuclear physics. The memoirs of science are rich with instances of this internationality and its personal manifestations. For example, Arthur Compton, who received the Nobel Prize in physics in 1927, relates that in 1931, when he visited Rome, he and his wife were received at the Fermis' home. "How little we dreamed," Compton wrote, "that ten years later we should be welcoming them to Chicago to share with us a task of nuclear physics research whose purpose would be nothing less than to preserve the world's safety and freedom." [2] Furthermore, no segment of science was in the midst of such great and exciting intellectual achievement as was nuclear physics. An epoch of experimentation and theorizing was culminating in new and revolutionary understanding of the most fundamental building block of matter, the atom; and this understanding held forth the possibility that incredible amounts of energy might be extracted from the atom. The rapid pace, the excitement, the scientific enchantment of this work sent its devotees back and forth across the Atlantic for consultations and deliberations, and, it is important to note, the ripeness of nuclear physics for historic achievements served to draw the very brightest young men into apprenticeship in the field. The transatlantic ferment also had other effects. It served to give American nuclear physicists a firsthand acquaintanceship with the violent anti-intellectualism of the Nazis and the Italian Fascists. Enrico Fermi, perhaps the most idolized and beloved member of the international science community, chose not to return to Italy after going to Stockholm in 1938 to receive the Nobel Prize. Einstein, the most towering and revered theoretical figure in the twentieth century scientific revolution, similarly left Germany. They and hundreds of other refugee scientists

---

whelming that scientists confronted with the exigencies of national need have reacted as much as other patriotic citizens, professional and nonprofessional." I believe that Frutkin applies a useful corrective to the gushiness of those who contend that science is an international force that exists above and beyond national politics. The fact is that when the shooting starts, scientists almost invariably go to work to enhance the destructive power of their own nation. Nevertheless, science is the most international of pursuits, with the possible exception of diplomacy and banking, and as long as they are not engaged in trying to kill each other's fellow countrymen, scientists of all nationalities have a great deal to do with each other and get along remarkably well.

took up residence in the United States, where their fellow scientists felt blessed to furnish them haven.* Those physicists who knew Germany—and few had not included German laboratories in their training—were appalled by the Nazis' destruction of the German university system. By 1937, 40 percent of German professors had been dismissed for political reasons.[3]

Thus, while isolationism flourished in the United States, the scientific community, and particularly the physicists, had ample opportunity to become acquainted with fascism's impact. And all this happened at a time when nuclear physics was advancing to an understanding that held forth the promise of releasing the energy of the atom—an area, as physicists knew, in which Germany held a commanding position despite its anti-intellectual depredations.

In 1939, Edwin M. McMillan, a young physicist who was later to receive the Nobel Prize, was in the office of Ernest O. Lawrence, the inventor of the cyclotron, who was to be honored that year with the Nobel Prize. They were listening to a broadcast of a speech by Hitler. McMillan recalls, "Lawrence turned to me and said, 'We've got to stop that man.' " Throughout the nation, and throughout the scientific community, similar sentiments were increasingly felt and expressed. But probably nowhere were they stronger than among the nuclear physicists, the internationalists whose work had opened the possibility that the atom might become a weapon of war.

* It is by no means unusual for persons in one land to extend assistance when crisis or catastrophe, of whatever origin, afflicts persons in another. It can be argued, however, that the scientific community is more sensitively attuned to such situations than are other professional groups. In 1966, for example, when Argentine universities were brutally attacked by a military government that had taken office through a coup, the American scientific community swiftly organized to offer assistance. The National Academy of Sciences sought places in American universities for ousted and self-exiled Argentine faculty members and students. And the Academy, through its Foreign Office, collected the signatures of influential American scientists and educators for a letter to the president of the military regime. The letter, in part, warned that an exodus of scholars would be detrimental to Argentine economic development and added that "it would be in the [Argentine] national interest to seek the means to retain all professors who have submitted their resignations and to ensure that their future work is facilitated through evidence of greater respect for their profession." (*Science*, November 25, 1966, p. 992.) It is difficult to determine the effects of this appeal, but the regime was subsequently reported to have tempered its harshness toward the universities.

✿   ✿   ✿

Late in 1938, Otto Hahn and Fritz Strassman, in the Kaiser Wilhelm Institute for Chemistry, in Berlin, directed a neutron bombardment at a uranium target and detected a radioactive barium isotope among the by-products. Hahn communicated his findings to an Austrian colleague, Lise Meitner, who, along with her nephew, Otto Frisch, concluded that the barium by-product indicated that a new type of nuclear reaction had taken place—fission. In the international community of nuclear researchers, the findings were instantly recognized to have profound implications. On January 26, 1939, Enrico Fermi and Neils Bohr publicly presented a review of the latest developments at the Fifth Washington Conference on Theoretical Physics. The revelations of the Washington meeting immediately attracted the attention in this country of what was then probably the most outstanding government laboratory in the physical sciences, the Naval Research Laboratory. NRL's preeminence was no accident. The Navy, because of its dependence upon astronomy for navigation and meteorology for weather forecasting, had, of necessity, become more involved with science and technology than had the other military services or other government departments. Furthermore, the technology of the industrial revolution was peculiarly adaptable to marine warfare. Between the end of the Civil War and the beginning of World War I, there were major refinements in the technology of land warfare, but nothing of the magnitude that transformed the major navies of the world. The battle at the Marne was technologically no more than a massive extrapolation of the trench warfare of the Virginia peninsula. But between 1865 and 1914, navies had rapidly evolved from sail and primitive steam vessels to heavily armored oceangoing fleets, as well as long-range submarines carrying highly sophisticated armament. Thus, the Navy had a powerful incentive to keep abreast of science and technology, and to explore their implications for naval warfare. For example, as early as 1922, at NRL's predecessor, the U. S. Naval Aircraft Research Laboratory, Navy researchers were engaged in studies that suggested what was eventually to become radar.[4]

The birth of this nation's effort to build the atomic bomb is generally traced to a letter that Albert Einstein wrote to Franklin Roosevelt on August 2, 1939, to alert the government to the military potential of the atom. The letter merits a place in the history of the bomb project, but, in fact, it was only part—a rather small part—of a complexity of events that both preceded and followed its delivery to the President. These events are worth examining in some detail, for they tell a good deal about the formative days of the contemporary politics of science.*

The January 1939 discussion of fission by Bohr and Fermi quickly aroused the Navy's interest in atomic energy as a source of propulsive power for naval vessels. This interest apparently was intensified by what appears to have been the first attempt by members of the scientific community to alert the government to the possibility of developing atomic energy for military purposes. On March 16, 1939, three months after Bohr and Fermi spoke in Washington and six months before the Einstein letter to Roosevelt, George B. Pegram, dean of the graduate faculties at Columbia University, wrote to the Navy that preliminary investigation suggested that uranium might "liberate a million times as much energy per pound as any known explosive." [5] Fermi, in discussions with the Navy, indicated that he was pessimistic about the possibility of extracting this energy. The conclusion of this approach was that the Columbia group and the Navy would keep in touch. A few days later, however, the Naval Research Laboratory offered the Carnegie Institution of Washington $1500 to help investigate the power potential of uranium. But at this point, there intruded the traditional aloofness of science and government. Carnegie at first agreed to accept the sum, but later, for reasons of internal policy, rejected the offer. As a privately endowed research institution strongly imbued with the traditions of science's aloofness from government, Carnegie was prepared to use its own resources for cooperating with government, but it did not wish to risk the possibility of contamination through government subsidy.†

These inconclusive contacts were frustrating to other scientists, espe-

---

* The widely held interpretation of the role of the Einstein letter is reflected in an account published in the *New York Times* upon his death in 1955: "By one of the strangest ironies in history, it was Einstein, the outstanding pacifist of his age, who initiated the move that started the United States on the project that led to the development of the atomic bomb. . . . On reading the Einstein letter, President Roosevelt called in the late Brig. Gen. Edwin M. Watson, his military secretary, known as Pa Watson. 'Pa,' said the President, 'this requires action.' With these four words, the atomic bomb project, which was to cost $2,000,000,000, got underway. It is doubtful whether any other man except Einstein could have moved the President. . . ." (*New York Times,* April 19, 1955.)

† I am indebted to Caryl P. Haskins, president of the Carnegie Institution of Washington, for making available to me materials in the Institution's archives. Much of the material contained in these archives is beyond the scope of this particular study, but my impression is that the Carnegie papers constitute a virtually untouched treasure trove for students of the postwar period in general and of science and government in particular.

cially those who had fled from Europe and who were fearful that Germany, with its great scientific capability and aggressive intentions, might be first to exploit the atom. Among them was Leo Szilard, a member of the remarkable group of Hungarian scientific émigrés that included Eugene P. Wigner, Edward J. Teller, and John von Neumann. Upon learning in June 1939 that government restrictions made it difficult for the Naval Research Laboratory to enter into cooperative arrangements with non-government laboratories, Szilard conferred with Wigner and later with Einstein. Having perceived the military significance of the atom, the three, later joined by Teller, then took their case directly to Roosevelt, in the form of a letter signed by Einstein, who, to the lay world, was the most prestigious figure of science. It is worth noting that at the time they addressed themselves to the President, the distance between science and government was still so great that no explicit request was made for a government-supported research program. Rather, Roosevelt was asked to entrust a confidant with the task of alerting government departments to developments in nuclear research as well as to recommending steps for securing an adequate supply of uranium, the principal source of which was then the Belgian Congo. As for the actual research, it was suggested that the person selected by the President should attempt "To speed up the experimental work, which is at present being carried on within the limits of budgets of university laboratories, by providing funds, if such funds be required, through his contacts with private persons who are willing to make contributions for this cause, and perhaps also by obtaining the cooperation of industrial laboratories which have the necessary equipment." [6]

The attempts by Pegram, Fermi, and Einstein to alert their government to the potential significance of the results of fundamental research is a landmark in the history of science and government. But the approach was from the working level, not the institutional leadership, of science; and it came at a time when the barriers between academic science and government, though receding, were nevertheless still formidable. Roosevelt responded affirmatively to the Einstein appeal, but neither imaginatively nor forcefully. Reaching into the long-neglected government technical service, he appointed an Advisory Committee on Uranium. The chairman was Lyman J. Briggs, director of the National Bureau of Standards, an agency that was traditionally a starveling in the federal hierarchy. Serving with Briggs were two ordnance specialists, one from the Army, one from the Navy. The Army man, how-

ever, did not see too much promise in atomic energy, and on one occasion recited to the committee the Army's traditional litany that wars
are won by men and morale, not by weapons.[7] The committee sought
the advice of knowledgeable scientists who were interested in promoting
atomic research, including Pegram, Fermi, Szilard, and Wigner, but
thinking small was an inbred trait of the government service, and the
Briggs committee, though maintaining contact with the work at Columbia and elsewhere, was opposed to any large-scale efforts to achieve a
chain reaction until preliminary research had further progressed.[8] On
November 1, 1939, the committee reported to the President that military applications of atomic energy "must be regarded only as possibilities."[9] For further exploring these possibilities, government funds
totaling $6000 were made available for the next year. The most power-
ful technological innovation of all time was not to come to fruition
through grass-roots agitation and low-level government study.

Einstein and company could awaken interest, but what was needed were
men who could break traditions and move institutions. In 1940, while
the Briggs committee cautiously studied the atom, such a group was
coalescing. Its leader was one of the great and in many respects, tragic
heroes of the nation's history, Vannevar Bush, an autocratic Yankee
electrical engineer and pioneer in computer design who had left the vice
presidency of MIT in 1933 to become president of the Carnegie Institution of Washington. The Institution, with its Department of Terrestrial Magnetism, its Geophysical Laboratory, and other research
centers, was among the most respected nonacademic research organizations in the country. Prior to coming to Carnegie, Bush had successfully
practiced science on the border line of basic and applied research, applying the theories of Norbert Wiener to the construction of machines
for mathematical analysis. Thus, by virtue of professional achievement
and office he had entree to both the academic and industrial research
communities. If there was a scientific establishment in 1940, it could
be said that Bush was high in its ruling echelon. Furthermore, Bush
constituted the strongest personal link then existing between the federal
government and academic science. For several years before moving
to Carnegie, he had been a member of the National Advisory Committee for Aeronautics (NACA), which occupied a unique position as
an adviser and dispenser of federal funds for aeronautical research in
government and nongovernment institutions. In 1939, he became chair-

man of NACA. As an inventor and holder of patents, he had a long-standing interest in the patent system, which brought him into contact with Roosevelt's confidant, Harry Hopkins, who sought his advice on a proposal to establish a National Inventor's Council. Bush also had a long interest in the applications of science and technology to war, an interest that went back to his participation in antisubmarine warfare research in World War I. In 1925, as a naval reserve officer, he began to explore the problems of applying computers to the control of anti-aircraft fire. As chairman of a committee under the unfortunate Science Advisory Board, Bush was aware of the traditions that would have to be overcome if science and technology were to come into partnership with government. But now there was a major difference. The utility of science and technology in the economic crisis of the early New Deal was never clear-cut. In fact, it was even argued in the early 1930's that science and technology had contributed to the crisis by outdistancing economic and social understanding.[10] But in mid-1940, with radar just beginning to play a critical and dramatic role in the aerial battles over Britain, there was no difficulty in demonstrating that science and technology were indispensable ingredients of modern warfare. In every respect, Bush was tailor-made for exploiting the new atmosphere and for mobilizing and employing the technical community for war. "I was located in Washington, I knew government, and I knew the ropes," he recalled years later. "And I could see that the United States was asleep on the technical end." Bush himself, however, was wide awake. Though personal participation in advanced research was long behind him, he was well equipped for precisely following the work of others. In 1940, when the United States government, through the Briggs committee, was working with the grand sum of $6000 to assess the military potential of the atom, Bush decided that he would seek funds from the Carnegie trustees to push work in this field. In some respects, he was going far out on a limb. On March 31, 1939, an eminently distinguished physicist on his staff had sent him a memorandum on research related to "speculations now current with regard to the possibility of obtaining 'atomic power' from uranium by a chain-reaction process due to neutrons." The memorandum concluded that the "measurements do not indicate that such a chain reaction is possible. . . . Even if we assume that our present measurements are wrong . . . it appears to us that there is no reason for any great excitement." Bush nevertheless pushed ahead on exploring the possibilities, despite another memorandum from the same scientist, dated May 7, 1940, which concluded, "The idea of using uranium fission for making bombs of enormous explosive power does not appear very practical at present."

In a career that had crisscrossed the scientific community, Bush, however, had virtually automatic access to whatever was going on on the frontiers of research. On May 9, 1940, Karl Compton, president of MIT, wrote him a confidential letter describing a conversation with Alfred L. Loomis, a distinguished physicist and attorney who presided over his own research organization, The Loomis Institute for Scientific Research in Tuxedo Park, New York. Compton reported: "As far back as the first suggestion that uranium fission might have a very significant industrial or particularly military significance, he (Loomis) has been very close to the groups involved in this work. . . . It is now clear that the German scientists are concentrating major efforts on this problem at the Kaiser-Wilhelm Institute. It also appears clear that uranium 235 could be a tremendously powerful war weapon if it could be secured in substantial quantity and a fair degree of purity. Alfred makes the very pertinent suggestion that we really ought to get together some of the most competent men in the field to analyze the possibilities in the situation and be ready to proceed actively if a promising program develops." Compton went on to note that "George Kistiakowsky is intensely interested in the subject (as a weapon which must not be allowed to develop first in Nazi hands) and he is also making an independent study and will report to Alfred Loomis. The latter is also going to consult Pegram again in a few days and plans to get in touch with you." [11]

On May 23, 1940, the Carnegie Executive Committee provided Bush with $20,000 to be used at his discretion, "for a defense research project concerning uranium fission."

In view of the current balance between federal and private support of research, there was to follow from these events a stunning bit of irony. For the Briggs committee—which was created in response to Einstein's appeal—was so strapped for funds, and so enmeshed in red tape, that it soon turned to Carnegie for financial assistance in carrying out its mission. On June 24, 1940, for example, Briggs wrote to Bush, requesting that the Carnegie Institution make available $1000 "to cover traveling expenses of members of the Committee, telephone calls and similar items. It is extremely difficult," he noted, "to provide such expenses from Government funds, particularly if the meeting is not held in Washington." He also asked that $2000 be provided to Professor

Jesse W. Beams, of the University of Virginia, "to cover summer salaries of his associates and assistants and to provide for the purchase of necessary materials. This will permit the work on the centrifuge to be pushed vigorously until such time as the Government is in a position to give it added support." [12] On August 24, Briggs wrote to Bush to recommend that $500 be made available to pay for equipment and supplies for A. K. Brewer, a Department of Agriculture physicist, to study the separation of uranium isotopes. Briggs concluded his letter with, "I wish to say at this time that the funds you have made available to the committee have been of tremendous value to us to get started on a number of important leads and in providing for committee travel. I am watching these outlays carefully." [13]

Meanwhile, through his professional and personal acquaintanceships, Bush had set about enlisting the institutional leadership of science for an appeal to the White House to begin mobilizing science and technology for war. His colleagues in this were James B. Conant, president of Harvard; Compton at MIT, and Frank B. Jewett, president of both the National Academy of Sciences and the Bell Telephone Laboratories —one of the few industrial laboratories that was accorded the respect of the academic scientific community.* Their objective was to establish an administrative framework that would enable the scientific community to avoid its doleful experiences of World War I, when scientists who wished to contribute to the war effort usually found themselves in uniform and subordinate to scientifically illiterate military men. In that conflict, a presidential executive order had established the National Research Council (NRC) as the operating arm of the National Academy of Sciences to direct the employment of science for war. But NRC could neither operate nor contract for the operation of laboratories; rather, it was a clearinghouse, adviser, and would-be coordinator of research that was principally carried on in military laboratories. For many sci-

---

* Bush had been in touch with Jewett and Conant, as well as with Compton, during the early deliberations over whether to proceed with an atomic development program. Among these four, as among many others involved in the research and deliberations, it had been agreed that secrecy would prevail. Nevertheless, on May 27, 1940, the New York *Herald Tribune* carried a conspicuous and lengthy report, headlined, "Science Seeks Final Solution of How To Use Atomic Power." The article reported on work at various laboratories and even noted that "Fears have been expressed that Germany might be ahead of the rest of the world. . . ." Jewett sent a copy of the article to Bush, with the following note: "The boys will talk and write (and generally advertise)." (Jewett to Bush, Carnegie archives.)

entists, World War I provided a stultifying experience that reinforced their doubts about government's ability to employ science effectively.

Through his relationship with Hopkins, Bush was admitted to see Roosevelt early in June 1940, when the *blitzkrieg* was rolling across France. The President, eager for the support of any segment of American society that would help pull the nation away from its isolationist complacency, responded, per the design of the Bush-Conant-Compton-Jewett alliance, with an executive order that, in a sense, constituted the birthday of the great postwar involvement of science and government. Dated June 27, 1940, the order established the legal mechanism for what American science had long found unattainable: government support *and* autonomy. The administrative device for achieving this was through the creation of a National Defense Research Committee (NDRC), which one year later, along with the medical arm of the wartime effort, the Committee on Medical Research, became part of a new and far more comprehensive organization, the Office of Scientific Research and Development (OSRD). Directed by Bush and dominated by civilian scientists, OSRD, with Conant as the number-two man, comprised a civilian-controlled preserve, reporting directly to the President, working toward military objectives, in close liaison with the military, but independent of military control. By Bush's design, and with the approval of the President, OSRD was not simply to fill orders for the military; rather, it was to be a source of weapons creativity, unencumbered by what technically untutored military men conceived to be useful and possible. For the future of American science, however, the key point was that the work of OSRD could be contracted to university laboratories, on a flexibly drawn contractual basis, designed to assuage the scientific community's traditional fear of government interference with scientific independence. As described by Irvin Stewart, who served as deputy director of OSRD, "The heart of the contract problem was to reconcile the need of the scientist for complete freedom with assurances that Government funds would not be improperly expended. . . . The performance clause [of the contract] was a relatively simple provision. The contractor agreed to conduct studies and experimental investigations in connection with a given problem and to make a final report of his findings and conclusions . . . by a specified date. This clause was deliberately made flexible in order that the contractor would not be hampered in the details of the work he was to perform. The objective was stated in general terms; no attempt was made to dictate the method of handling the problem." [14]

The scientists had triumphed. Though the urgencies of war circumscribed the research effort, and virtually eliminated fundamental research in favor of satisfying immediate needs, the very creation of OSRD was a political landmark for the nation's scientific enterprise. For the first time in the nation's history, substantial federal funds were going to university laboratories. Furthermore, while the contract was intended to reconcile freedom with accountability, it was clearly weighted toward freedom.

Now, with his line of authority running directly to the White House, one of Bush's first steps was to dispose of the still-floundering Briggs committee and get on with building a bomb.

NOTES

1. Sir Gavin de Beer, ed., *The Sciences Were Never ct War* (New York: Thomas Nelson and Sons, 1960), p. 26.
2. Arthur H. Compton, *Atomic Quest, A Personal Narrative* (New York: Oxford University Press, 1956), p. 16.
3. Samuel A. Goudsmit, *Alsos, The Story of a Mission* (New York: Henry Schuman, Inc., 1947), p. 191.
4. Robert M. Page, *The Origins of Radar* (Anchor Books; Garden City, N. Y.: Doubleday & Co., Inc., 1962), pp. 19–22.
5. Quoted in R. G. Hewlett and O. E. Anderson, *The New World, 1939/1946*, Vol. 1, *A History of the United States Atomic Energy Commission* (University Park, Pa.: Pennsylvania State University Press, 1962), p. 15.
6. Lewis L. Strauss, *Men and Decisions* (Garden City, N. Y.: Doubleday & Co., Inc., 1962), Appendix, p. 438.
7. Ralph E. Lapp, "The Einstein Letter That Started It All," *New York Times Magazine,* August 2, 1964.
8. Hewlett and Anderson, p. 23.
9. Arthur H. Compton, p. 28.
10. A. Hunter Dupree, *Science in the Federal Government* (Cambridge, Mass.: Belknap Press of Harvard University Press, 1957), p. 349.
11. Karl Compton to Bush, letter, Carnegie Institution archives.
12. Briggs to Bush, letter, Carnegie Institution archives.
13. *Ibid.*
14. *Organizing Scientific Research for War* (Boston: Little, Brown and Co., 1948), p. 19.

V

The Experience
of War

The failure of German nuclear physics during World War II can in large measure be attributed to the totalitarian climate in which it lived. . . . By putting politics first and science second, the Nazis contributed greatly to the deterioration of German scientific teaching and research. To be a good Nazi was not necessarily synonymous with being a good scientist and we can learn from their mistake to leave science to the scientists.

SAMUEL A. GOUDSMIT, in
Alsos, The Story of a Mission, 1947

Since the politics of science are intimately connected to the conditions that scientists deem desirable for carrying on research, the genealogy of the politics trails back to the beginnings of science. In all times, those who have systematically sought after an understanding of the physical universe have had to arrive at some sort of *modus vivendi* with the surrounding mass and with each other. For the American scientific community, as we have seen, the understanding that prevailed with the nation at large, right up to World War II, was predicated on a kind of mutual aloofness accompanied by a genteel poverty for science. This aloofness was eroding, rather rapidly, as the war approached, and it was forever gone when the war ended. But the war was to bring not simply an acceleration of the pace. Rather, in a period of months, patterns that had more or less prevailed for 150 years were cast away and replaced; with astonishing speed, the aloofness was destroyed, and science, fed by government subsidy, was linked to government purpose. For the scientific community, this was an intoxicating and traumatic experience, an experience that was formative of personal and institutional relationships that today, a quarter of a century later, still endure to a very large extent in the politics of science. But to understand these politics it is necessary to recognize that the war's impact on pure science must not be read only in terms of the new institutional arrangements or the ever-growing budgets that were to sustain the newly formed partnership of science and government. These were the material, easily measurable consequences of the war-born relationship. Far more important, I believe, were the psychological consequences, for the American scientific community emerged from the war with the conviction that its contributions to victory were, first of all, indispensable. But even more important, in terms of planning how the prewar orphan was to be treated in the postwar period, the scientific community held that its contributions would have been unattainable if the management of the most significant wartime science and technology had not been left to scientists. Since history permits no reruns, the thesis cannot be tested. But the evidence is persuasive.

After the Briggs committee had inconclusively studied the feasibility of atomic energy for over a year, Bush vigorously stepped in and took over, employing an administrative device that swiftly cut the ground out from under the presidentially appointed committee chairman. Since the National Academy of Sciences was formally the scientific adviser to the federal government, Bush prevailed upon Briggs to request the Academy to review the activities of the committee. In large part, Bush was inspired by the concerns that Ernest O. Lawrence, inventor of the cyclotron, had conveyed both to him and to Karl Compton about the slow pace of the Briggs committee's proceedings.[1] There *were* great uncertainties about the possibility of developing atomic energy for use in the war; some of the most important exploratory research was under way but yet to be completed, and Bush himself felt some sympathy for Briggs' cautious approach. But Briggs was more awed by the uncertainties than inspired by the possibilities, and he accordingly welcomed a review by the prestigious Academy. The review committee was placed under chairmanship of Karl Compton's brother Arthur, chairman of the physics department at the University of Chicago. Compton's conclusion, as stated fifteen years later in his memoirs, was that "not a single member of the Briggs committee really believed that uranium fission would become of critical importance in the war then in progress. This committee had been considering these possibilities for a year and a half. Its members were chosen from those well-qualified as to first-hand knowledge. No one felt keenly enough its possible contribution to the war effort to move away from another field of study into that related to fission research. The central concern of those promoting the uranium program was not defense but a source of power in peacetime." [2] At the time of the Academy review, major scientific problems still remained to be resolved. For example, how much fissionable material was necessary for a critical mass? Could the means be developed for producing and extracting the material? But by the fall of 1941, Compton's committee was speaking of the possibility of a "fission bomb of superlatively destructive power." [3] Research in progress at various institutions, including Columbia and the Carnegie Institution, and in Great Britain, encouraged the Academy committee to be bold where the Briggs committee had earlier been cautious. It is possible, though doubtful, that this research would have inspired the Briggs committee to the very conclusions reached by Compton's group. But in terms of the state of mind of the leaders of the scientific community at the end of the war, their perception of the experience is perhaps more important than the realities of the experience. The atomic bomb project began in earnest almost immediately following the reviews conducted by Compton's committee. And when this project culminated in the war's greatest technological triumph, Bush, Conant, the Comptons, and their other colleagues, virtually all with backgrounds in academic science and engineering, could reasonably conclude that it was scientists who were free of government, but free to employ government funds, who had brought about the triumph.

The wartime research and development effort, in and out of OSRD, was vast and unparalleled in terms of manpower, expenditures, and achievement. But throughout the war, science run by scientists drew and fulfilled the most difficult assignments. Aversion to military and industrial control was, and remains, an enormously powerful force throughout the scientific community. The Los Alamos Scientific Laboratory, which fabricated and tested the atomic bomb, was originally to be a military laboratory. But prospective recruits from academic science rebelled when they learned they were to work under the military. As described in the official history of the Atomic Energy Commission, their feelings were that "a military laboratory could accomplish nothing significant. Differences in rank between commissioned and civilian researchers would breed friction and bring on a collapse of morale. They believed that military organization would introduce a dangerous rigidity. Would not an Army officer find it difficult to be wrong, to change a decision? What assurance was there that he would act on scientific grounds? The laboratory must be civilian to retain scientific autonomy." [4] The civilians prevailed, and the Los Alamos laboratory was organized as an adjunct of the University of California. The scientific staff, drawn mainly from academic science, preferred to retain the status of university employees, though the connection between Los Alamos and the University of California was nominal.

But even when university status was provided for the great wartime laboratories, there were scientists who felt that any concession to the ways of the military was unbearable. E. U. Condon, later to become the director of the National Bureau of Standards, described himself as unable to serve at Los Alamos. Following a period there, he wrote to J. Robert Oppenheimer, the director, that "I found the extreme concern with security morbidly depressing—especially the discussion about censoring mail and telephone calls, the possible militarization and complete isolation of the personnel from the outside world. I know that before long such concerns would make me be so depressed as to be of little if any value. . . . I can say that I was so shocked that I could hardly believe my ears when General Groves, director of the Manhattan Project, undertook to reprove us, though he did so with exquisite tact and courtesy, for a discussion which you had concerning an important technical question. . . ." [5]

Condon's abhorrence of military control and security was widely shared in the scientific community, though the vast majority of scientists had no doubt that some measure of security was a necessary evil in wartime. In fact, many of the most fervent believers in the necessity for scientific freedom, among them Leo Szilard, had voluntarily agreed among themselves in 1940 to withhold from open publication research papers in nuclear physics that might be of assistance to the Germans.[6] Nevertheless, the scientists' abhorrence of military domination contributed to their insistence upon some form of civilian control of the most technologically advanced, and, as it turned out, the most militarily significant laboratories. The National Defense Research Committee established the Radiation Laboratory—the "Rad Lab"—under MIT, as the great center for radar research. The self-detonating proximity fuse, which was little short of the atomic bomb as a technical achievement, and infinitely more useful in fighting the war, was initially researched under contract between the NDRC and the Carnegie Institution; in 1942, Johns Hopkins took over the contract, and out of this arrangement there grew what is now the Applied Physics Laboratory. The

Metallurgical Laboratory, where the first controlled chain reaction was achieved, was operated by the University of Chicago. A similar reliance upon civilian-controlled university research prevailed through other operations of the Office of Scientific Research and Development, often to the annoyance of the technical branches of the military services, which interpreted the use of university laboratories as a self-serving device conceived by academic scientists who had gained the favor of the White House. As the official history of OSRD frankly acknowledges, in referring to the relationship between the old-time Army Chemical Warfare Service and the independent, politically insulated wartime civilian research organization: "It was inevitable that NDRC would be regarded with suspicion since it was an independent agency free from Army control. On its side the NDRC was bound to regard the Army as set in its ways, as uncooperative, and as unwilling to accept new developments which it had not originated itself. The competition which existed between the two organizations led inevitably to duplication of effort and wastage of time." Time and the urgencies of war rubbed down the rough edges, and by the end of the war "there existed the closest cooperation, intermingling of personnel, and mutual tolerance, even though not always mutual affection." [7] But those who had long suffered in the ranks of the government's traditionally neglected technical services were not without grounds for resentment as they observed the affluent, confident Bush enterprise. The civilians regarded government science as inferior, plodding science. As the physicist Luis Alvarez put it, "At Berkeley before the war, when we thought of government research, we thought about the nearby Department of Agriculture laboratory, and that was everything that was wrong with science." * Furthermore, to the resentment of the military services, the OSRD leadership quite sensibly concluded that the wartime research program could be meshed with one of the principal objectives of academic research—the training of new scientists.[8] Perpetuation of the profession being one of the irreducible values of the scientific community, the scientific leadership saw no reason why the long-time well-being of the scientific com-

---

* Agricultural research is frequently a whipping boy of those who have made their careers in the physical sciences. One reason is that the peer and project systems, based on professional assessment of scientific achievement and potential, have generally governed the allocation of federal money in the physical sciences. In the agricultural sciences, however, Congress long ago established a formula system that provides each state with a share of the available funds—regardless of the capability of the researchers who receive the funds.

munity could not be reconciled with the short-term requirements of war. Accordingly, Bush and his colleagues argued that scientists and science students were more valuable to the nation in school or in the laboratory than in the infantry. "In 1914 Britain sent into the front lines one of the greatest of modern physicists," Bush wrote to the Secretary of War in 1944, in a plea for deferment of scientists. "He had revealed for the first time some of the secrets of the nucleus of the atom, and he was known throughout the world as one of the geniuses of his day. His name was Mosely. He was soon killed in action. . . .

"This example," Bush continued, "was repeated, with men of lesser calibre, many times, the war effort of Britain in the first war was severely crippled by this act of folly, and the nation finally learned a lesson. In this present war, Britain has utilized its young scientists well." [9] The United States was, in fact, also employing its scientists quite well, but as a matter of policy, deferred them from conscription on an individual basis, rather than providing a blanket deferral which might be politically difficult to defend to the parents of nonscientists. Nevertheless, Selective Service, hungry for manpower, did not always agree with the leaders of the scientific community, an experience which further convinced Bush and his colleagues that science is too important, too delicate, and too poorly appreciated by laymen to be under the sway of nonscientists. The available evidence suggests, however, that a disparity existed between the scientists' experience with Selective Service and their perception of the experience. According to Irvin Stewart, who served as deputy director of OSRD, deferments were requested for 9725 employees of OSRD contractors; of these, only 63 were inducted into military service.[10] Nevertheless, shortly after the war, Bush told a Senate committee that "it is my personal opinion, by and large, that throughout the war we took [into military service] more men of special training, scientists and engineers, out of industry and the laboratories than was wise for the best prosecution of the war." [11]

While the World War II research effort was continental in scope (OSRD, financially a junior partner in the overall effort, placed contracts in excess of $1 million each with fifty universities and industrial firms), it was relatively concentrated in scientific quality and in the political experience that it provided for scientists. This concentration was to have great significance in the postwar politics of science, for to a remarkable extent, the alumni of a relatively few wartime research laboratories were to predominate in the councils that were formed to link science to government. It was at Los Alamos and the MIT Radiation Laboratory that the United States gathered many of its most distinguished practitioners in physics and adjacent fields to concentrate their abilities on the most difficult technical problems of the war. And if we look back from the perspective of today, we find that an extraordinary proportion of the dominant figures throughout the postwar politics of science were academic scientists, primarily basic researchers, who served in either or both of those laboratories during World War II. For example, since 1951, seven men have served as science adviser to the president or chairman of the Executive Office Science Advisory Committee (SAC) that preceded the establishment of the advisory position. The first, Oliver Buckley, of Bell Telephone Laboratories, was the only one of the seven who was not a Los Alamos and/or Rad Lab alumnus. He occupied the White House position briefly, and at a time when the role and status of the SAC were uncertain and the chairmanship was a part-time position. Following his resignation, Buckley played virtually no role in science and government affairs. However, the six who followed him were all drawn from academic science, engineering, or administration, all were associated during World War II with one or both of the laboratories, and both before and after their periods of White House service, were deeply involved in the politics of science. Buckley was succeeded in the SAC chairmanship by Lee DuBridge, a physicist who was director of the Rad Lab, and later president of the California Institute of Technology. Next came I. I. Rabi, the Nobel laureate physicist, who served as associate director of the Rad Lab. Following Sputnik, when the full-time position of presidential special assistant for science and technology was created, it was filled by James R. Killian, Jr., an administrator, who was executive vice president of MIT during the war, and a member of the Rad Lab steering committee. Killian was succeeded by George B. Kistiakowsky, a physical chemist at Harvard, who headed the explosives division at Los Alamos. Next came Jerome B. Wiesner, an electronics engineer who served at both laboratories and later became director of the postwar successor to the Rad Lab. And Wiesner was followed by Donald F. Hornig, a physical chemist at Princeton, who served as a group leader at Los Alamos.

World War I has been referred to as a chemists' war; World War II as a physicists' war, though physics, in this context, must be broadly defined as including disciplinary neighbors and offspring, such as physical chemistry and electronics. With the atomic bomb and radar being conceived of as the most strategically vital tasks of the war, Bush and his colleagues aimed at concentrating the most creative scientific and technical talent in these fields. The atom bomb obviously called for the services of the nuclear physicists and physical chemists who, throughout the previous decade, had been close to the fundamental research that led the way to atomic energy. At least until the midpoint of the war, when the project moved from a laboratory to an industrial scale, it was these disciplines that predominated. And to the very end, it was Los Alamos, under the direction of a theoretical physicist, Oppenheimer, that served as the focal point for the nationwide effort to build the bomb for use in the war. The other great concentration of physicists was, of course, in the MIT Radiation Laboratory, the center for radar research. It was here—in contrast to the Manhattan Project—that the wartime research reached the ultimate of science governed by scientists. Early in the war, the bomb project was placed under the jurisdiction of the War Department. Though Los Alamos and the Metallurgical Laboratory retained civilian leadership and university affiliations, there was an Army general in overall command, Leslie R. Groves, who was quickly to become regarded as the *bête noire* of the atomic scientists. To the extent that a militarily uninhibited atmosphere prevailed in the atomic project, it was as the consequence of struggle between the civilian and the military, since Groves and the academics building the bomb might well have been thrown together by some mischief-loving force interested in producing maximum friction. The atomic scientists, fresh from the congenial anarchy of the campus, found themselves under a chief who was later to write: "Compartmentalization of knowledge, to me, was the very heart of security. My rule was simple and not capable of misinterpretation—each man should know everything he needed to know to do his job and nothing else." [12] By the traditions of science, no concept could be more absurd or offensive. Since the objective of the researcher was to learn the unknown, how was "everything he needed to know" to be delineated beforehand? The answer, of course, was that it could not be, and Groves' compartmentalization schemes were inevitably stretched or pushed aside by the pressures of getting on with the job. Groves later contended that he took this in stride. But one wonders. "While I was always on the other side of the fence, I was never surprised when one of them [the scientists] broke the rules. For example, I got through talking with Niels Bohr on the train going to Los Angeles . . . I think I talked to him about twelve hours straight on what he was not to say. Certain things he was not to talk about out there. He got out there and within five minutes after his arrival he was saying everything he promised he would not say." [13]

The reality of the atomic job was that the military's security phobia was a nuisance, and often an extremely annoying one. But if the success of the bomb project is any measure, it was not a disabling one. In January 1942, with the principal Manhattan Project centers still to be built and enormous scientific and technical uncertainties still to be resolved, Arthur Compton outlined a schedule that called for a completed atomic bomb by January 1945.[14] The first bomb was tested July 16, 1945. The realities of the conflicts between the atomic scientists and the military administrators are difficult to assess, but the perceptions that the scientists derived from the experience contributed mightily to their conviction that science fares best in the hands of scientists.

While the atomic scientists struggled with the military, and accumulated real and imagined grievances, the radar scientists operated from an enclave that, considering the circumstances of war, was as ideal as could be hoped for. The atomic scientists, with the exception of the Metallurgical Laboratory staff in Chicago, were dispersed to odd and often barren corners of the nation, the Los Alamos mesa, the wilds of Oak Ridge, Tennessee, and Hanford, Washington. But radar research was centered in the nation's academic mecca, Cambridge, Massachusetts. Furthermore, its genesis was entirely different from the bomb project's. The civilian scientists had found it necessary to convince the political and military councils of the potential of atomic energy; the project had had a slow and uncertain start, and the mysteries of the atom served to inspire the military, the political leadership in Washington, and the scientific leadership itself to attempt to impose a degree of security commensurate with what was to be the ultimate weapon. Radar had no such background. The military services had been puttering with it for years before the war; its utility was forcefully demonstrated by the British experience, and the Army and Navy felt an acute need for expert help if their forces were to be equipped with this new and indispensable device of war. One consequence of this was that immediately after NDRC was established, the armed services asked it to begin fundamental investigations into radar.[15] Compton promptly established what came to be known as the Microwave Committee, which, following consultations with British physicists who had been pioneering in the field, decided that it would be necessary to establish a major new laboratory exclusively devoted to radar research.

Who was to staff it, and where was it to be located? That physicists and their disciplinary cousins should predominate in the atomic bomb project was inevitable. The bomb was fundamental nuclear physics translated into nuclear technology, and there was no doubt as to where the necessary expertise was to be found. But who were the specialists in radar? Radio engineers, electrical engineers? They had been working with radar for twenty years, and though inadequacy of funds partially accounted for the still-primitive state of the technology, a more pertinent factor was inadequacy in the understanding of the fundamental principles governing microwave transmission. When fear of German air attack led the British Air Ministry in the mid-1930's to seek new means for detecting hostile aircraft, the British, with a well-rooted working relationship between science and government, had turned to their basic physicists. As Sir Robert Watson-Watt, popularly known as the "father" of radar, has written: "Radiolocation was the direct, but in no wise the predestined and inevitable, fruit of pure research. Its beginnings lay in the work of those who laboured to understand more of the things that happened in the earth's atmosphere. Its later developments, and much of its technique at all times, were due to those who sought the inner secrets of the structure of matter. . . . These researchers were State-aided, but most generously and most lightly State-controlled." [16] It was a group of British physicists, headed by N. L. Oliphant, of Birmingham, who produced what was to become the heart of radar—the resonant cavity magnetron. In the late summer of 1940, a British scientific mission, headed by Sir Henry Tizard, presented one of these revolutionary devices to Bush's newly formed organization. OSRD's official history refers to it as "the most valuable cargo ever brought to our shores." [17] But the British also brought something less tangible but no less important: their experience in putting basic scientists to work for war and especially at putting physicists to work on radar.

"We worked from the British experience," said I. I. Rabi, who served as associate director of the Rad Lab. It was obvious at the outset, he explained, that nobody knew anything about radar, but the physicists had "the intellectual mobility" to find out. "They were light on their feet," say Luis Alvarez, who served at the Rad Lab and in the Manhattan Project. "They knew something about electronics because of their work with accelerators, but the real reason was that they were the best people and they were adaptable to anything." Rabi adds, "The people at the Rad Lab didn't know a damn thing about radar. I was in charge of a magnetron group and I'd never seen one. Well, I thought, I'll go around to MIT and ask some of the electrical engineers. After talking to them, I could see that they didn't know anything either. . . . So we started absolutely fresh and designed magnetrons." An immodest, arrogant assessment of the wartime experience? Perhaps. But the physicists were "light on their feet," for in short order they achieved stupendous success in work distantly related to their peacetime occupations.

The Rad Lab experience was heady, intense, and formative of judgments and values that were to endure throughout the postwar period. For example, Edward M. Purcell, a Harvard Nobel laureate in physics, who served as a group leader at the Rad Lab, recalls, "Rabi had a very interesting role at the Radiation Laboratory which I have never seen adequately described . . . but it is often said that Rabi helped us keep our eye on the ball. Rabi is and was a very great man for listening to a discussion of policy or technical policy which gets off into details and pulling the whole thing together and saying what we really have to do and what is really wrong with it. He was in kind of an elder statesman role, even in those days."

If American scientists needed any reinforcement for their conviction that science managed by scientists produced the best wartime results, it was amply provided when victory permitted them to scrutinize the manner in which the Axis powers had employed their scientific and technical resources. A study mission that rushed into Germany on the heels, and occasionally ahead, of the Allied armies found that it was not a failure of espionage that had prevented the Allies from learning of the German atomic bomb project; rather it was simply that the relatively small German nuclear project that did exist was bogged down in blind alleys, at least three years, and probably a good deal more, behind the American program. The head of the mission, Samuel A. Goudsmit, concluded that the cause of the German failure was Nazi meddling with science. "The only branch of science in which the Germans outdistanced us," he observed, "was aerodynamics and this was because Goering, who was boss of the show, cut across party lines and gave the scientists a free hand and unlimited funds." [18] *

Further examination of Germany's wartime research revealed that in 1943, when Allied airborne radar, developed by the Radiation Laboratory in cooperation with the British, was at last winning the Battle of the Atlantic, Grand Admiral Doenitz wrote: "For some months past, the enemy has rendered the U-boat war ineffective. He has achieved this objective, not through superior tactics or strategy, but through superiority in the field of science; this finds its expression in the modern battle weapon—detection. By this means he has torn our sole offensive weapon in the war against the Anglo-Saxons from our hands. It is essential to victory that we make good our scientific disparity and thereby restore to the U-boat its fighting qualities." [19]

Repeatedly, the investigation of the performance of the German research effort added to the sense of triumph that American scientists felt for having had the wisdom and political opportunity to have science governed by scientists. Germany, expecting a short conflict, did not attempt to follow the Bush prescription for employing science and technology for war, and deliberately did not mobilize its scientific and technical resources. Stated OSRD's official history: The Germans felt that "if the war was to be a short one, they needed no great organization for the development of new weapons, no mobilization of their vast resources of scientific personnel, no experiments to ensure the scientist freedom to create within the meshes of their huge military system. . . . German war research in the early years of the war was therefore with few exceptions confined to the laboratories operated by the armed forces and those of the war industries. . . . By Hitler's orders, basic research on radar, for example, was stopped in 1940, and was not renewed until

---

* An alternative theory offered by some German scientists is that, for humanitarian reasons, they deliberately dragged their heels on nuclear research. See Robert Jungk, *Brighter Than a Thousand Suns, A Personal History of the Atomic Scientists* (New York, Harcourt, Brace and Co., 1958), pp. 91–104, for a discussion of this thesis. The evidence is mixed. For example, Heisenberg, the distinguished German physicist, and his associates tended to cast doubt on the feasibility of a bomb project, but, as we have seen, so did some American scientists, and there is no evidence that they were motivated by anything but technical considerations. Furthermore, there is no doubt that for purely political reasons, the German program was placed in the hands of a number of second-raters who would never have been admitted to positions of similar responsibility in the American program. It is worth considering, too, that there is no evidence that the nonnuclear aspects of the German wartime research effort were ever impeded by humanitarian concerns.

1942. The heavy hand of bureaucracy forced industrial as well as government laboratories to concentrate their efforts on the improvement and testing of existing weapons." [20]

In contrast, the Bush-led enterprise flourished and produced magnificently. From the politically insulated laboratories there came the atomic bomb and the proximity fuse—which the Germans also tried but failed to master. While German radar research was stagnating, the Rad Lab was pioneering in the development of novel and superior microwave equipment. Even that ancient instrument of war, the bow and arrow, was subjected to scientifically rigorous improvement, under the direction of Paul Klopsteg, a physicist and archery enthusiast, who took on the task of devising new and highly lethal bow-and-arrow-type devices for the use of commandoes and agents of the Office of Strategic Services.*

The scientists did not confine themselves to the laboratory, but often developed their weapons by going to the theaters of operation to assess the needs, returning to the laboratory to design the weapons, and then returning to the battlefield to provide instruction on the employment of their creations. This procedure, which also had roots in the British radar experience, conformed to the advice of Sir Isaac Newton, who wrote: "If instead of sending the observations of seamen to able mathematicians at land, the land would send mathematicians to sea, it would signify much more to the improvement of navigation. . . ." [21]

When the military was slow to acknowledge the wisdom of the civilians, vindication was all the sweeter after these newcomers to military affairs were demonstrated to be right. Bush writes that the DUKW, the tracked amphibious vehicle that played a major part in clearing the Pacific, "came into existence because of the vision of a small group of civilian engineers, plus the encouragement of unconventional generals with a flair for pioneering. In fact, there was probably more obtuse resistance to this device than to any other in the war." [22]

But obtuse resistance, however intense at the beginning of the war, was becoming a rarity as the military came to realize the immense value of their newfound colleagues in war. Following the American

---

* Ten years after the war, Klopsteg's murderous designs were still classified as confidential by the Defense Department. Bureaucratic lethargy was probably the reason, but the Department chose to use the explanation that it believed it undesirable for these designs to fall into the hands of teen-agers or criminals. However, after a spate of newspaper publicity, the designs were released.

reconquest of the Philippines, as forces were being mounted for the projected invasion of Japan, MacArthur's headquarters gave whole-hearted approval to plans for OSRD to establish in Manila a scientific headquarters consisting of a senior consulting scientific staff, a pool of scientific specialists, plus laboratory facilities "suitable for emergency work such as might be possible in the theater." [23] The war ended before these plans were fully put into effect, but at the cessation of hostilities, sixty scientists were working for MacArthur's command, and another two hundred were lined up to join them. The arrangements for this scientific force were initiated by Alan T. Waterman, deputy chief of OSRD, who later was to become the founding director of the National Science Foundation.

It is difficult to exaggerate the effects of these wartime experiences on the psyche of the American scientific community. Since the work of OSRD was climaxed by victory in the greatest of wars, events took on the effect of ratifying the wisdom of the manner in which OSRD operated.* OSRD could look back over its incredible five-year history and pick out examples of brilliant performance and foresight. In 1941, for example, Bush wrote to Warren Weaver, chief of the OSRD applied mathematics panel, "We ought courageously to look forward and plan the ideal [aircraft] gun control. I believe that we ought not only to plan it, but that we ought to build it. It ought to be so far in advance that it will at the present time be somewhat unrelated to what is feasible, but it ought to have those features which we believe will prove essential as air warfare continues." [24]

But Bush, the seasoned engineer and administrator, knew that knowledge by itself is inert, that some mechanism must exist to turn knowledge into utility. In things small and large he always sought to provide for that transition. Thus, he decreed that the use of OSRD personnel in operations analysis must be with the understanding "(a) that the

---

* There were a few dissenters, but their voices were poorly heard and the validity of their plaints is difficult to assess. For example, two months after the war ended, a group of Minnesota researchers declared, "When the true record is written, the waste, inefficiency, ignorance, and obtuseness in utilizing scientific knowledge in the recent war will be apparent to all." (*Hearings on Science Legislation,* Subcommittee of the Senate Committee on Military Affairs, 1945, p. 963.) The record, as written so far, fails to substantiate this doleful prophecy; nevertheless, it is not unlikely that the official OSRD histories, as well as the memoirs of OSRD's figures, tend to pass over whatever blemishes did exist.

officer to whom they are detailed definitely wants them; (b) that they be allowed access to such information as they may need for their work; (c) that they be allowed reasonable freedom as to the way in which they do their work; and (d) that they be responsible to the Commanding Officer and make their reports and recommendations to him, distribution of such reports beyond the Command to be subject to his approval." [25]

When the time came to proceed from experiment to production in the atomic bomb project, some of the younger scientists cockily assumed that they could easily command the transition from the laboratory to an industrial operation. The official history of the AEC reports, "Compton had already encountered 'near rebellion' among his staff at the Metallurgical Laboratory when he suggested that an industrial organization familiar with large-scale operations be given the task of constructing and operating the production plants." [26] Bush knew the range and limitations of scientific expertise. After the scientists had fairly well established that a bomb could be built, DuPont and other industrial organizations were brought in to build it. But science took credit for the bomb, on the grounds that its contribution was the indispensable ingredient.

The experience of war was intoxicating for many members of the American scientific community. They went into the war convinced that science is best run by scientists. They emerged with this conviction reinforced by experience, their own selective vision, and the awe and gratitude of the scientifically illiterate lay world. As one distinguished physicist noted: "Suddenly physicists were exhibited as lions at Washington tea parties, were invited to conventions of social scientists, where their opinions on society were respectfully listened to by lifelong experts in the field, attended conventions of religious orders and discoursed on theology, were asked to endorse plans for world government, and to give simplified lectures on the nucleus to Congressional committees." [27]

Measured in terms of its acceptance by the nation, the American scientific community had come an immense distance since the New Deal, just a decade before, had turned down its pitiful plea for financial assistance.

Obviously the government would now support science. Science was necessary to national well-being, and there was no alternative to government support. But, in peacetime, could science expect support as well as independence? On the basis of their wartime experience, Bush and his colleagues saw no reason why not.

NOTES

1. R. G. Hewlett and O. E. Anderson, *The New World, 1939/46,* Vol. 1, *A History of the United States Atomic Energy Commission* (University Park, Pa.: Pennsylvania State University Press, 1962), p. 36.
2. *Atomic Quest, A Personal Narrative* (New York: Oxford University Press, 1956), p. 46.
3. Hewlett and Anderson, p. 47.
4. *Ibid.,* p. 231.
5. Quoted in Leslie R. Groves, *Now It Can Be Told* (New York: Harper and Brothers, 1962), p. 429.
6. *Atomic Energy, Hearings,* Special Committee on Atomic Energy, United States Senate, 1945, p. 289.
7. James P. Baxter 3d, *Scientists Against Time* (Boston: Little, Brown and Co., 1946), p. 279.
8. Baxter, p. 278.
9. Baxter, p. 128.
10. Irvin Stewart, *Organizing Scientific Research for War* (Boston: Little, Brown and Co., 1948), p. 274.
11. *Hearings on Science Legislation,* Subcommittee of the Senate Committee on Military Affairs, 1945, p. 212.
12. Groves, p. 140.
13. *In the Matter of J. Robert Oppenheimer,* Transcript of Hearing before Personnel Security Board, Atomic Energy Commission, 1954, p. 166.
14. Hewlett and Anderson, p. 55.
15. Baxter, p. 140.
16. "Radar in War and Peace," *Nature,* Vol. 156 (September 15, 1945), p. 321.
17. Baxter, p. 142.
18. *Alsos, The Story of a Mission* (New York: Henry Schuman, Inc., 1947), p. xi.
19. Quoted in Baxter, p. 46.
20. Baxter, p. 7–8.
21. H. W. Turnbull, *The Correspondence of Isaac Newton* (New York: Cambridge University Press, 1961), Vol. III, p. 364.
22. Vannevar Bush, *Modern Arms and Free Men* (New York: Simon & Schuster, Inc., 1949), p. 35.
23. Baxter, p. 416.
24. Baxter, p. 220.
25. Bush, memorandum to OSRD Advisory Council, August 14, 1943; quoted in Baxter, p. 411.
26. Hewlett and Anderson, p. 90.
27. Samuel K. Allison, "The State of Physics, or the Perils of Being Important," *Bulletin of the Atomic Scientists,* January 1950, p. 3.

VI
Meshing the Incompatible

... if the direction comes from the top down ... scientists feel that they are just being pawns, that they are just being paid to do a job, [and] that isn't the way that you are going to get a big scientific program. . . . The scientists, after all, know more about science than anybody else in America.

> IRVING LANGMUIR, Associate Director,
> General Electric Laboratory,
> before Subcommittee of Senate Committee
> on Military Affairs, 1945

I regret very much ... that the subject of what we do about research gets into the position of the scientists telling us how to organize the Government, because I think that is one area in which their competence doesn't meet the situation.

> HAROLD D. SMITH, Director, Bureau of The Budget,
> before Subcommittee of Senate Committee
> on Military Affairs, 1945

World War II shattered the long-dominant European scientific community. But the historically penurious American scientific community, though almost wholly diverted to war work, was invigorated and enriched by the war. Europe lost Fermi and Einstein; the United States gained them. Many of the workshops of European science were weakened or destroyed by war, but the workshops of American science were strengthened by the forced-draft application of science and technology to war. After the liberation, while European science faced the task of reconstruction, the American scientific community was going through what was probably the first of the postwar revolutions of rising expectations. The war demonstrated not only that big organization, big equipment, and generous funding were compatible with the creative process, but that with war-born instruments and technology ready to be applied to basic research, bigness had become indispensable in many

97

fields of research, especially physics. In the prevailing understanding, technology was the offspring of fundamental knowledge—knowledge led to devices and gadgets; but, in fact, the instruments of war that were developed through fundamental knowledge could just as well serve to produce new fundamental knowledge. The financially threadbare basic research of the 1930's had produced the fundamental knowledge that was embodied in radar and in the nuclear reactors that created fissionable materials for atomic weaponry. Now the vacuum and electronics technology that had been developed for radar opened possibilities for a new generation of particle accelerators to explore deeper into the atom; similarly, the reactors that were hurriedly built for war posed exciting possibilities for a new generation of equipment useful in a broad range of scientific research. Leland J. Haworth, later to become director of the Brookhaven National Laboratory, an AEC commissioner, and then director of the National Science Foundation, recalled that before the war, he and fellow students at the University of Wisconsin scavenged parts from gasoline pumps to piece together an accelerator. "Those were indeed hard times. . . . We were so poor that we could not afford a forty-dollar variable transformer. . . ." [1] Luis Alvarez recalled similar prewar experiences at Berkeley. "In peacetime, when we finished with a part, we carefully unsoldered the connections and put everything back on the shelf." During the war, both served at the Rad Lab, where money was the least consideration in getting the job done. "I'll never forget," said Alvarez, "the time when we were setting up the Radiation Laboratory, and the boys went to the radio shops in Boston and ordered everything they wanted."

In planning the plutonium project, Arthur H. Compton proposed an initial six-month budget of some $300,000. "This amount seemed big to me," he later wrote, "accustomed as I was to work on research that needed not more than a few thousand dollars a year." [2] A survey covering the 1939–40 academic year in thirteen leading universities revealed that the highest total of departmental expenditures in physics for "direct operating expenses of research" (defined as including "equipment, apparatus, technical and research assistance, publishing cost associated with research, field trips, expeditions, etc.") was $39,000; in chemistry it was $73,000. [3] By contrast, at the end of the war, OSRD could report that it had awarded contracts totaling nearly $117 million to MIT, $83 million to Caltech, $31 million to Harvard, and $28 million to Columbia. [4] Virtually all of the money was for war-related technology, not for science. But scientists, along with others with technical competence, employed these unprecedentedly large funds to meet wartime needs, and in the process savored the experience of ample resources for the first time. The scientists did not wish to return to impoverishment. The condition and nature of the American economy dictated that government would have to be the principal source of support for their postwar ambitions. But how was this to be accomplished? What were to be the mechanisms for delivering support unaccompanied by the control or influence that science had long feared and resisted from any patron, but especially government?

In 1944, a number of developments led Bush to realize that it was time to lay plans for nourishing science when peace returned. The great bulk of the scientific community was too preoccupied with its wartime tasks to devote any extensive thought to the postwar financing and administration of research. But in 1944 the massive wartime research effort was already beginning to taper off at certain points, and there was restlessness in the ranks. At the Metallurgical Laboratory, in Chicago, the task of designing the plutonium production technique was nearing completion. The scientists there, unlike their colleagues at other installations in the far-flung Manhattan Project, were thus relieved of the onerous schedules that had bound them for nearly three years, and inevitably, their thoughts turned to the future. By all accounts, this did not happen to any significant extent at the Rad Lab, at Los Alamos, or any of the other research installations which operated full blast to the end of the war.* But at Chicago, the successful fulfillment of their assignment brought the scientists time to meditate, and their thoughts turned mainly to two matters: the moral and political implications of the weapon they were helping to build, and where the money would come from to exploit the exciting research possibilities that had been

* Academy President Frederick Seitz, who did research on various military projects while serving on the faculty of the Carnegie Institute of Technology through most of World War II, has written; "During the war relatively little thought was given by the universities to the influence of federal support on institutional policies, because most university staffs not only were absorbed in war activities but also felt that the flow of government contract money would stop promptly at the end of the war. Most of us hoped only that we would be permitted to keep some reasonable fraction of the equipment and related facilities that had been accumulated during the war." (Frederick Seitz, "The University: Independent Institution or Federal Satellite?" *Science and the University*, Boyd R. Keenan, ed. [New York: Columbia University Press, 1966], p. 151.)

opened by their wartime work. It was obvious that neither prewar politics nor prewar budgets would suffice in the nuclear age, but with the war still raging, how was a hearing to be obtained for a group of scientists who, because of the peculiarities of the division of labor, were almost uniquely moving toward idleness? The answer arrived at by the Chicago scientists was that noise and eloquence were the best available substitute for power. And soon they were to be inspired to direct a great deal of both toward Bush and the other wartime leaders.

At the beginning of 1944, the Metallurgical Laboratory was filled with rumors that by June, 90 percent of the personnel would be released. Compton, fearful that his distinguished but restless staff would be lured away, sought General Groves' support for gradually moving the laboratory into a postwar research program. But Groves, under ever-increasing pressure from Washington to complete the bomb, was not disposed to divert resources from the prime objective. He responded that he approved of Compton's proposals in principle but forbade any commitments to research of long-range significance. There was to be no major construction or detailed design efforts. The laboratory was to go on a standby basis around the beginning of September, Groves decreed, and he also advised Compton to prepare plans for a 25 to 75 percent reduction of staff. Probably nothing could have been more shattering to Compton, who coyly acknowledged: "Looking to the future, it appeared that there would be substantial advantage to the University of Chicago if some of this research could be continued." [5]

As one of the architects of scientific planning for the war, Compton had located the plutonium project at Chicago, despite his own recognition "that the main focus of these [research] activities at this time was in Columbia and Princeton and that establishing the center at Chicago would require the moving not only of many research men but also of several carloads of equipment." [6] But the decision was Compton's to make. Ernest O. Lawrence thought "the tempo of the University of Chicago is too slow," and that the plutonium project should go to his own laboratory, at Berkeley.[7] On top of the history of Compton's favoring of Chicago, reports now came to Bush that Compton was trying to persuade Fermi to leave Columbia permanently, and had offered him a professorship along with the suggestion that Fermi might become director of the nuclear research laboratory that had grown out of the war effort at nearby Argonne.[8] Thus, even with the war still raging, there was jockeying for position in the postwar period. To buttress

Chicago's position, Compton appointed a special study committee, headed by Zay Jeffries, a General Electric metallurgist and executive who served at Chicago as a consultant. With Fermi and other distinguished members of the laboratory serving on the committee, it produced "Prospectus on Nucleonics," a 65-page glowing report which spelled out the scientific, industrial, and military potential that could be expected if the federal government provided substantial financing for research in nuclear energy and electronics.*

The Jeffries report also touched upon another matter that was troubling the Chicago scientists—national security in the atomic age. International control of atomic energy, it argued, was essential to avoid wars infinitely more destructive than the conflict that was then raging. Safety was not to be found in secrecy or in an attempted moratorium on research, but in "setting up an international administration with police powers which can effectively control at least the means of nucleonic warfare."

In forwarding the Jeffries report to General Groves, Colonel K. D. Nichols, deputy head of the Manhattan Project, sent a brief covering memorandum that stated, "Your particular attention is directed to recommendation E on page 65"—which read: "Enlightenment of public opinion on the scope and significance of nucleonics should start as soon as possible to bring about realization of the dangers for world security caused by new scientific and technical developments, and to prepare for decisions which will have to be taken to meet this danger." [9]

The seeds of dissension between science and the military had been sown long before. Now, on the issue of how atomic energy was to be managed in the postwar period, they were beginning to sprout.

For Bush and the other scientist leaders, the agitation on politics and finances was a signal that planning for the postwar period must start at once. At Bush's request, Richard C. Tolman, one of General Groves'

---

* The Jeffries report was probably the first of a genre that was to become increasingly common throughout the postwar period: the brief for government support of a particular line of research or a scientific discipline, with the case stated in terms of the ripeness of the scientific opportunities and the probable payoffs in knowledge and utility. Twenty years later, under the auspices of the National Academy of Sciences, descendants of the Jeffries report were being produced by committees of earth-based astronomers, chemists, oceanographers, physicists, and biologists, each group stating the case for support of its field of research. The Jeffries report, which, as far as I know, has never been published in its entirety, is in the Manhattan Engineering District collection of the National Archives.

principal assistants, headed up a Committee on Postwar Policy that sought to develop comprehensive plans for the support of science. Bush assured Compton that the Chicago laboratory would not be permitted to disintegrate.* Meanwhile, there was a war to be won, and though the scientists might be concerned about how the politicians would treat them in the postwar period, the politicians who could decide the matter had their attention fixed on more urgent matters. Even in less turbulent circumstances, the response of the politicians might have been the same, for traditionally, one of the most grievous of science's political problems has been—and still remains—that few politicians share the scientists' vision of science's value and place in the nation's life.

While restlessness in the ranks provided Bush with an incentive to get on with postwar planning for science, an even greater motivation was provided by events on Capitol Hill. There, Senator Harley M. Kilgore, a West Virginia New Dealer, continually on the lookout for issues to mold to his populist vision, held the chairmanship of the Subcommittee on War Mobilization of the Committee on Military Affairs. The Senator, perhaps more than most legislators, hungered for problems to solve. (Upon his death in 1956, *The New York Times* observed: "There was hardly a major national problem during his years in office for which he did not offer his own special solution. The fact that his solutions were almost unanimously ignored by his colleagues never seemed to daunt him. He always offered more.) [10] Seeing how well science was serving the war, layman Kilgore—to the horror of scientist-engineer Bush—proceeded to draw plans for a great postwar collaboration of science and government, a matter which the Senator felt could be accomplished without any significant departures from the

---

* Events vindicated this assurance, for Chicago ultimately was to retain and build upon the research effort that started there during the war. But Groves refused to be stampeded by the Chicago petitioners. On August 21, 1945, for example, he wrote to Chicago Chancellor Robert Hutchins: "From the papers, I note that you are planning on establishing an institute which would include among its fields that of nuclear physics. I have had many discussions in the past with Dr. Arthur H. Compton, who has expressed the view that the University of Chicago should be a center of such research in the years to come, and that it should receive government owned equipment which has been in use at the Metallurgical Laboratory, including the Argonne installation, during the last few years. I feel it is incumbent upon me to inform you, in order that there may be no misunderstanding, that there can be no commitment with the University at this time on the part of the War Department in this request." (Groves to Hutchins, Manhattan Engineering District collection, National Archives.)

administrative procedures traditionally followed by government. There would, of course, have to be a new federal agency—a foundation of some sort—to support research, the Senator reasoned. But, as was the case with virtually all federal agencies, he felt, it should be headed by a presidential appointee. To the extent that the Senator recognized that basic science had peculiar vulnerabilities and sensitivities, he believed that these could be accommodated by providing a special board to serve as adviser to the presidentially appointed director. The board would consist of eight public members, presumably scientists and administrators, and nine government officials: the director, plus the heads or representatives of the War, Navy, Interior, Justice, Agriculture, Commerce, and Labor departments, and the Federal Security Agency. For Bush, the apostle of science governed by scientists, especially academic scientists, the administrative structure proposed by Kilgore was sufficient to inspire gross alarm, but the Senator also had other designs that perhaps were even more disturbing. Oblivious of the historical plight of basic research in the United States, and of the basic researchers' long quest for political recognition of the vulnerabilities of fundamental science, Kilgore threw together basic and developmental research. Under Kilgore's design, the new agency would be authorized to range from research in the basic sciences to research in "methods and processes beneficial to small business." [11] He thus ignored the prewar experience of the short-term quest for utility draining off support from the long-term and uncertain process of seeking fundamental knowledge. Furthermore, the populist layman architect of science and government proposed that patents resulting from federally financed research become public property, a proposal which Bush, who was acquainted with the problems of nursing scientific findings into marketable products, considered a blow to the prospects for expanding industrial research. Whatever the ideological merits, the fact was, Bush felt, that industry did far too little basic research, and could hardly be expected to enter into partnership with the federal government for the conduct of research if any competitor would be free to exploit the results. In Kilgore's view, the motivation for federal support of basic research was not "building up theoretical science just to build it up. The purpose is what has been the purpose of scientific research all the war through, and what the incentive for scientists is, to do something for the betterment of humanity" [12] —which was another way of asserting what had historically been the bane of basic research in the United States: emphasis on utility.

The Kilgore concept for supporting science "holds grave danger to the full development of science," Bush declared in an appearance before the Senator's subcommittee.[13] And twenty years later, Bush commented, "Kilgore's plan would have set up a gadgeteer's paradise." But what was most alarming to the wartime leader of American science was that Kilgore's prescription for the political linkage between science and government appeared to be in harmony with the thinking of some key administration figures.

By 1944, no life remained in the issue of whether government should support science. ". . . We are all agreed that scientific research and development must be undertaken into the indefinite future," said Representative Sterling Cole (R-NY), in a statement that typified congressional sentiment.[14] But he added what was to become the key issue between the sensitive prewar orphan and its newfound patron: "The question is how it can be done most effectively." As for the scientists, their prewar fears of government were intact. But these were overshadowed by economic realism and the wartime experience of science supported by but insulated from government. "I realized that science would fall flat on its face after the war if the government didn't keep up its support," said Bush. The issue was now the means toward an agreed-upon objective, but, very specifically, it came down to whether government would accord basic research a peacetime enclave in which it could operate with the generous, continuous support and freedom that had characterized OSRD.

Bush now resorted to the behind-the-scenes maneuvering that had served so well when he and his colleagues brought the scientific community to the service of government on the eve of United States entry into the war. At that time, following consultations with Harry Hopkins and a brief visit to Roosevelt, he had come away with instructions to prepare a letter in which the President directed him to establish NDRC. Now, Bush's first step, as he explained years later, was to inform Roosevelt of his misgivings for Kilgore's design. Shortly after this, he was to lay before Roosevelt a letter in which the President asked the director of OSRD to propose a series of recommendations for government's postwar relationship with science. Bearing the President's signature, under the date November 17, 1944, the letter, in effect, gave

Bush carte blanche to prepare a blueprint for continuing and extending the war-born relationship.*

Bush's prime concern was, of course, the historically tenuous position of academic basic research. But in addition to being a scientific administrator, Bush was also a political merchandiser. (Compton recalls in his memoirs that in 1941, when Bush and his colleagues decided to advise Roosevelt to proceed with the bomb project, "At the suggestion of Bush the cost estimate was played down, lest the government should be frightened off.") [15] With postwar planning for science now under way, Bush framed his charter from Roosevelt in terms of what science could do for government and society, not in terms of what the scientists were more deeply concerned about, namely, what government could do for science.

Stated the Bush-drafted presidential letter: "The information, the techniques, and the research experience developed by the Office of Scientific Research and Development and by the thousands of scientists in the universities and in private industry, should be used in the days of peace ahead for the improvement of the national health, the creation of new enterprises bringing new jobs, and the betterment of the national standard of living." [16] Specifically, the head of OSRD asked the President to direct him to make recommendations on four issues: (1) The release, consistent with military security, of scientific and technical information developed during the war. This was a reflection of Bush's concern that the Army would attempt to impose blanket secrecy on

---

* A differing version of the letter's genesis is offered in John Gunther's *Taken at the Flood* (New York: Harper and Brothers, 1960), a biography of Albert Lasker. According to Gunther, Lasker's wife, Mary, long a key figure in medical politics and related national politics, was appalled to learn that OSRD's Committee on Medical Research would be dissolved along with the parent organization at the end of the war. Gunther relates, "She talked to Anna Rosenberg, who said, 'Give me a memorandum, and I will take it to F.D.R. . . .' Roosevelt turned the matter over to Judge Rosenman, who promptly wrote a letter in F.D.R.'s name to Dr. Bush, directing him to make a report." (P. 318.) In a paper prepared for the Science and Public Policy Seminar of the Graduate School of Public Administration at Harvard, C. B. Baldwin, Jr., notes that, in response to an inquiry, Judge Rosenman said he had no recollection of the incident; Mrs. Rosenberg, however, said Gunther's description was accurate. (C. B. Baldwin, Jr., "Federal Support of Research by the National Institutes of Health and the National Science Foundation," unpublished paper, 1961.) What is certain is that the idea of postwar government support for science was in the air, and that Bush was a key figure in various people's deliberations on the subject.

wartime research developments, especially in the atomic energy field. (2) The organization of a medical research program, which Bush and company regarded as scientifically desirable and politically popular. (3) The provision of government assistance "to aid research activities by public and private organizations," the closest that the letter of direction came to the essential concern of the scientific leadership—support of basic research in the universities. And (4) the creation of "an effective program . . . for discovering and developing scientific talent in American youth"—a new venture for government support, but one that inevitably appealed to science's evangelical spirit. Nowhere in the Bush-Roosevelt letter, it is interesting to note, is there any explicit reference to basic research, which was to be the focal point of the ensuing report, *Science, The Endless Frontier*. It was the commitment to basic research, a commitment of religious intensity, that was driving the leadership, and a large portion of the scientific community, in its behavior toward the liaison with government. Even General Groves, who, for many scientists, personified the dangers inherent in government-subsidized and -directed research, understood the scientists' concern about and intense affection for basic research. In describing the steps to be taken to hold together the staffs of the military laboratories after the war, Groves stated that the scientists, in addition to their weapons work, "are going to be given the equipment and they are going to be given the time to engage in certain fundamental research. . . . That is part, as far as I am concerned, of their salary." [17]

Bush swiftly convened panels, largely composed of his OSRD colleagues, to answer the questions he had asked Roosevelt to ask him, and on July 5, 1945, he delivered his reply to Roosevelt's successor in the White House, Harry S Truman. The report was accompanied by a routine letter of transmittal, which described the genesis of the study and incidentally added, "It is clear from President Roosevelt's letter that in speaking of science he had in mind the natural sciences, including biology and medicine, and I have so interpreted his questions. Progress in other fields, such as the social sciences and the humanities, is likewise important; but the program presented in my report warrants immediate attention." A decade had passed since a committee of social scientists had helped scuttle the Science Advisory Board's proposal to single out the natural sciences for federal subsidization. Now, from among the entire array of scholarly disciplines, the physical scientists were the accredited envoys to the Washington councils, and it was their judgment that their own work should be given priority in the new partnership between knowledge and government.

How was basic science to be supported after the war? Bush's reply was that it should be supported through the creation of a political channel that would accommodate a flow of money to the practitioners of science, but screen out the possibility of control by those who provided the money. *Science, The Endless Frontier* duly paid respects to the importance of maintaining "responsibility to the people, through the President and Congress," [18] but it then went on to set forth an administrative formula that, in effect, constituted a design for support without control, for bestowing upon science a unique and privileged place in the public process—in sum, for science governed by scientists, and paid for by the public. Whereas Kilgore had proposed that the Foundation be headed by a presidential appointee who would be advised by a board composed of government officals and private citizens, Bush proposed that the Foundation be headed by a presidentially appointed board of approximately nine private citizens who would appoint and supervise the director. The difference was a critically important one. For, though the President could control the makeup of the board, and, with the Congress, the appropriation of funds to the Foundation, the proposed administrative structure was such that the Foundation would be only tenuously linked to executive control. In a showdown, the traditional mechanisms of government would, of course, prevail, since they controlled the purse. But government is not conducted by a process of showdowns between the White House and executive agencies; rather it is conducted by a continuing effort to orchestrate the federal establishment, to bring the bureaucracy into harmony with whatever design is held in the White House. But under the Bush design, the proposed foundation would be free from the normal processes of executive discipline. Whereas law, tradition, or simply general practice called for publicly supported activities to be examined annually through the preparation of the executive budget and its review before Congress, Bush argued that the peculiarities of basic research required special political recognition and budgetary treatment. "Basic research is a long-term process—it ceases to be basic if immediate results are expected on short-term support. Methods should therefore be found which will permit the agency to make commitments of funds from current appropriations for programs of five years' duration or longer." [19] Bush was correct. The uncertainties and complexities of basic research are not easily reconciled with the vagaries of the federal budgetary process. But the same can be said of agriculture, military planning, and slum clearance. Why should science be accorded a place of privilege that was denied other supplicants at the public treasury? Looking for peacetime precedents for the administrative design proposed for the Foundation, Bush observed that the "very successful pattern of organization of the National Advisory Committee for Aeronautics, which has promoted basic research on the problems of flight for the past thirty years, has been carefully considered in proposing the method of appointment of Members of the Foundation and in defining their responsibilities." [20] But, however carefully the NACA experience was considered, the fact was that NACA was a political freak, riding on the glamour of aviation to escape the prewar pattern of government relations with research. Furthermore, nine of NACA's seventeen members were government officials, an example reflected in Kilgore's proposals but one that Bush had no desire to follow.

Such was the formula advocated by Bush, whose experience, personal ties, and concerns were closest to those of the now-triumphant physical scientists. But what of other fields of scholarship, particularly the social and medical sciences? How did they fit into Bush's design for the postwar collaboration of science and government?

What must first of all be observed about the social sciences is that until their relatively recent involvement with computers and large-scale surveys, their costs were trivial in comparison to most of the physical sciences. The expense of one of the major postwar particle accelerators, for example, could probably have underwritten years of the ongoing work as well as the aspirations of all of American sociology and behavioral psychology. Thus, the social scientists were not spurred by the economic considerations that caused their colleagues in the physical sciences to look to Washington. Furthermore, while atomic science and chemistry were apolitical, morally neutral, nonpartisan, and perhaps most important of all, unintelligible to laymen, the social sciences were, by their nature, frequently concerned with subjects that aroused strong feelings: race, economic policy, educational techniques, sexual mores— matters on which everyone feels he has a right to vie with the experts. Now on the brink of obtaining the support that had eluded the physical sciences through all of the nation's history, the statesmen of science were not about to jeopardize their chances by bringing the controversial social sciences into the discussion.

Among the advisory groups established by Bush to assist in the preparation of his report was the Committee on the Discovery and Development of Scientific Talent, chaired by Henry Allen Moe, secretary general of the Guggenheim Foundation and a lawyer by training. In his report to Bush, Moe noted that Roosevelt's letter ordering the study "refers to science as the word is commonly understood, or, more technically described, to science now within the purview of the National Academy of Sciences . . ." Therefore, Moe continued, his committee would limit its recommendations to the physical and natural sciences. But then he went on to argue that the "statesmanship of science, however, requires that science be concerned with more than science. . . . We could not suggest to you a program which would siphon into science and technology a disproportionately large share of the Nation's highest abilities, without doing harm to the Nation, nor, indeed, without crippling science." [21] In his report to the President, Bush referred to Moe's argument, and observed, "It would be folly to set up a program under which research in the natural sciences and medicine was expanded at the cost of the social sciences, humanities, and other studies so essential to national well-being." [22] Folly or not, when it came to detailing the activities that would be supported by the Foundation, the social sciences were nowhere among them. When the issue of the social sciences came up before a House committee considering the proposed Foundation, Representative Clarence Brown remarked, "There is a sort of antipathy against the social sciences." Replied Isaiah Bowman, a geographer who was president of Johns Hopkins University and former vice president of the Academy, "Your remarks . . . are . . . a summary of the views of most of the scientists who testified before the Senate committee." [23] Brown declared that support of the social sciences would result in "a lot of short-haired women and long-haired men messing into everybody's personal affairs." Oppenheimer, Conant, and a few of the other wartime leaders expressed support for including the social sciences in the Foundation. And apparently the rank and file of the scientific community shared this view. In 1948 *Fortune* polled some four thousand scientists on a variety of questions, including: "Do you think the social sciences should share in any disbursement of federal funds for research?" The answers were overwhelmingly affirmative, 81 percent in the case of academic scientists, 83 percent among those in government employ, and 76 percent among industrial scientists.[24] But the prevailing view among the leadership was that, at least initially, the physical and natural sciences should be exclusive beneficiaries of federal assistance for science. "Indeed," said Homer W. Smith, professor of physiology at the New York University School of Medicine, "the presence of the social sciences in the legislation would give rise to problems and difficulties which in our opinion would greatly prejudice the foundation." [25]

Kilgore reported: ". . . I have asked nine eminent economists to give me the name of a really authoritative, basic work on the fundamentals of economics. All of them say, 'There is no such book.' " [26] *

As far as the medical sciences were concerned, they did not initially share the scale of financial ambitions of their colleagues in the physical sciences, nor, for that matter did they have any desire to be administratively cast with them. Historically, a variety of factors had encouraged the medical sciences to develop apart from other fields of research. The required proximity of university medical schools to hospitals often resulted in these schools actually being remote from the main university campus, with little contact taking place between medical and nonmedical members of the faculty. During the war, OSRD's medical activities were compartmentalized in the Committee on Medical Research, which, though administratively an equal of Conant's National Defense Research Committee, had little occasion to become involved in the strategic and political deliberations that brought the physical scientists into close and regular contact with high level military and political circles. Furthermore, medical research had its own peculiar ideology, mystique, and economics. The pure physical and natural sciences could exist intellectually and economically apart from involvement in the affairs of society. Einstein, Fermi, Bohr, Lawrence were not unconcerned with man's fate—but the fundamental science they practiced was at best distantly related to the human condition. By contrast, until

---

* Twenty years later, the President's Science Advisory Committee was yet to include its first social scientist, a fact which one member, Edward Purcell, the Nobel laureate, physicist, Rad Lab alumnus from Harvard, explains as follows: "It would be very helpful in some cases to have an economist in there. But in social science you dip into a very broad spectrum of stuff. An individual social scientist may be a very good man, but he can't bring social science to bear on these problems as we can bring physics to bear on a technical problem. . . . I have heard arguments between an economist and a physicist, and the physicist says, 'Tell us one theorem that you fellows have that isn't trivial that you all agree on.' "

a great burgeoning of federal support for fundamental research in the life sciences in the mid-1950's, most medical research was in one way or another related to problems of disease, rather than a quest for understanding of fundamental biological processes. The research was often carried on either by practicing physicians or by those who had drifted away from practice, but in either case, the medical research community was close to, and shared much of the social and political conservatism that characterized the medical profession.

Against this background, it is not surprising that the medical advisory committee for the Bush report, composed mainly of medical school administrators and researchers, excelled the other disciplines in the quest for support and independence, to the point of strongly recommending that medical research be independent not only of government, but of the other sciences, as well. ". . . The Federal agency concerned with medical research should be created *de novo*," the committee declared, "and be independent of all existing agencies, none of which is sufficiently free of specialization of interest. . . ." [27] The committee spoke admiringly of the British Medical Research Council, an administratively autonomous body which dispenses government funds, adding, apparently for the peace of mind of the American medical profession, that the Council "has no connection with any system of medical care or health insurance." [28] Finally, while reciting the standard litany that money must be given without strings—and for periods up to ten years—the medical men displayed great caution toward the volume of money. Five to seven million dollars annually would suffice "in the immediate postwar period," they said, to which they appended a warning of "the ease with which the quality of the work can be lowered by encouraging men to undertake research who are inadequately prepared or unfitted for the task." [29] Bush, in his report to the President, said: "The amount which can be effectively spent for medical research in the first year should not exceed 5 million dollars. After a program is underway perhaps 20 million dollars a year can be spent effectively." [30] He ignored the medical men's recommendation that they be provided with a separate agency.

At the end of the war, the leaders of American science saw an unprecedented opportunity to gain the support that had long eluded their profession. Bush, the statesman, tempered the demands of some of his colleagues for the scientific community to be given political carte blanche. Moe's advisory committee on the Discovery and Development of Scientific Talent flatly proposed, for example, that science students in the Armed Services "should be ordered to duty in the United States for fully independent, integrated scientific study . . . and no desire of a commanding officer to retain a potential scientist for his usefulness on the spot should be allowed to interfere with the operation of the policy." [31] The proposal, it might be recalled, was made in June 1945, when forces were being assembled for the projected invasion of Japan. In his final report, Bush softened this recommendation by suggesting that the return of science students to their studies should be "consistent with current discharge plans" and "as soon as militarily possible." [32] But, outside of avoiding extreme demands for the political independence of science, Bush was not inclined to trim his prescription for government's postwar relationship with the scientific community. He was prepared to go a long way, in fact, not only to assure that academic science would have government support without government control, but also that academic science would have a powerful voice in the conduct of the government's own research organizations. "A permanent Science Advisory Board should be created," he recommended, "to consult with these scientific bureaus and to advise the executive and legislative branches of Government as to the policies and budgets of Government agencies engaged in scientific work. This board should be composed of disinterested scientists who have no connection with the affairs of any Government agency." [33] Bush had insisted that scientists in government employ should have little or no voice in the affairs of the proposed foundation; on the other hand, he deemed it desirable that nongovernment scientists should be placed in a position of influence over the policies and budgets of the government's research. When viewed against the background of American government tradition and practice, the prescription was brazen and outrageous. But it was natural, predictable, and perhaps inevitable when viewed against the traditions and practices of science, the painful indifference which science experienced before the war, and the scientists' intoxication by wartime success. The difficulty, however, was that the American government would not buy it.

A congressman taking his first look at the scientific community, as many of them did in the early postwar discussions, could easily conclude that science's ways of doing business were curious, contradiction-ridden, anarchic—but effective. J. Robert Oppenheimer, whose wartime leadership of the Los Alamos laboratory made him the nation's scientific folk hero, argued before respectful legislators that basic research was a delicate, unpredictable enterprise whose success depended upon simply leaving the researcher alone to follow his own curiosity. The researcher must not only be free from the untutored intrusions of government bookkeepers looking after the use of government funds; he must also be free from intrusions by his own scientific colleagues. "Let me ask you a question along that line," said Kilgore. "The director of the laboratory in a university, or the dean of the school, always has an idea of a particular field that his people are particularly adapted to work in; isn't that right?" Replied Oppenheimer, "It should not be so, though it sometimes is." To which he added cryptically, "A good physics department is one in which you can't find out who is the big shot." [34] Oppenheimer went on to explain "half of the work of physics and most of the work of fundamental physics is not to measure the quantities that are known to exist, but to find out what quantities do exist. This is, what kind of language, what kind of concepts correspond to the realities of the world. . . . If you really know what it is you have to do, if you can give it a name and say, 'We don't know whether it is 10 or 20, but we would like to measure it,' then you are doing something which increases knowledge very much but which doesn't qualitatively increase knowledge." A senator from Arkansas, J. William Fulbright, responded. "That was a fine explanation, but . . ." At which point, Senator Warren G. Magnuson of Washington interrupted to remark to Oppenheimer, "He doesn't understand you either." [35]

The scientists might plead for the sovereignty of the individual researcher in their quest to make science politically insulated, but in arguing that a board rather than a presidential appointee should govern the Foundation, Karl Compton declared, "By long experience I have come to have great faith in the combined judgment, knowledge, and wisdom of a small competent group, greater faith than in the ultimate decision by one individual." [36] James Conant added, "I feel it is most fortunate that in my university, I have little or no power to act alone." [37] The wartime leaders saw no paradox in their positions, and, in a sense, there was no paradox. Fearful of the possibility of political domination of the proposed foundation, they sought protection in collective leadership drawn from the scientific and academic communities. When many led, none could dominate, and thus the independence and freedom of the individual researcher could be assured. But for politicians bred on the traditions of strict accountability for government funds, the doctrine set forth by the scientists was troublesome.

The men who served in the executive and legislative branches of government required no convincing that science and government now needed each other. But as they pursued the task of determining *how* the heretofore privately supported laboratory was to become a financial ward of government, the behavior of the scientific community contributed to a great puzzlement. The scientists clearly hungered for federal funds to support the costly work they wished to pursue in the postwar period. But it was not simply money that they sought. Perhaps more than any other supplicants for public funds, they were concerned about the *conditions* that would accompany the money. Government support of basic research in the universities obviously could not be patterned after government dealing with industry, which would contract to perform a specified piece of work in return for a specified sum. The scientists and engineers who spoke for basic research had persuasively argued that basic research was a quest for the unknown and therefore could not be subjected to the time clock or other conventional criteria of performance. But the scientific leadership appeared to be intractable in its insistence upon exemption from the traditional accountability procedures, and it was especially averse toward procedures that would permit laymen to judge the worthwhileness of basic research. To a large extent, however, the scientists and the politicians were talking past each other. Francis G. Blake, dean of the Yale School of Medicine, could assert (and quite correctly) that the value of the penicillin mold might not have been recognized "had Fleming not had the freedom to putter away with this curiosity to his heart's content." [38] The implication was that scientific "puttering" could not long endure exposure to the will and impatience of the lay public and its elected leaders. But governmental traditionalists, while almost unanimously conceding the importance and peculiar requirements of basic research, were either reluctant or outrightly opposed to bestowing subsidy and sovereignty upon science. "There is always the disposition," declared Interior Secretary Harold Ickes, "whenever a new activity of Government is proposed for outsiders to rush in and say, 'We are the only people competent to do this work. We are not bureaucrats,' although some of them hope they will become so. 'We are not politicians. You can trust us to lift this up to a higher plane and keep it there.' " Concluded Ickes: "To me that is all bunk." [39]

Nevertheless, by pounding incessantly upon the concept that the uniqueness of science demanded unique political treatment, the scientists were getting through to many members of Congress. "When I was directing the research work of students in my days at Princeton University," Karl Compton explained, "I always used to tell them that if the results of a thesis problem could be foreseen at its beginning it was not worth working at." [40] Representative of those whose sentiments were tinged by these propositions, Fulbright, in a colloquy with Oppenheimer, expressed his doubts about the wisdom of Kilgore's formulation: "I am bothered by the feeling that since you don't know what you are doing [in the conduct of basic research] . . . and you can't account for it in the sense that this is going to take so many thousands of dollars, we run a great risk in giving this supervision to a Government bureau, which, under our old traditional system, always is supposed to set up a budget and say that so much is for nuclear, and so much is for this and that. . . . The danger of the [Kilgore] bill is that it seems to give such wide powers, and there is the implication that this [presidentially appointed] Administrator is going to tell [the scientists] what to do, that 'You go out and find some new kind of this or that,' which as I see from what you say, is contrary to the whole spirit of research. . . . The first thing you know, they will be demanding of the scientists, 'What are you going to spend this on?' and 'Did we get value received?' and so on. That is the danger of governmental control." [41]

In opposition to the Kilgore bill, the wartime leadership of the scientific community gave its support to a bill introduced by Senator Magnuson, which reflected Bush's design for a politically insulated enclave from which scientists could dispense government funds with minimal interference from or accountability to government. Under the Magnuson bill, the director of the Foundation would be selected by a presidentially appointed board which would be chosen on the "basis of demonstrated capacity for the job and not on an ex officio basis." Government officials thus would not be automatically placed on the board, and by any reasonable expectation, the board would become a creature of the scientific and academic communities. Finally, the Magnuson bill, unlike the Kilgore bill, did not stir up the issues of patents or small business.[42] Clearly, it was the legislative embodiment of the deep-felt ideology of the leadership of American science. But, in whatever form, the proposal for a new government agency to support basic research was soon to be overshadowed and confused by an altogether extraneous issue that split and scattered the energies of the scientific community—the peacetime control of atomic energy. The scientists were learning that the affairs of government can be infinitely more complex than the affairs of science.

The issue of postwar control of atomic energy placed Bush and his colleagues in a perplexing position, and spread confusion around their carefully devised plans for postwar support of basic research. In political circles, they were the most technically knowledgeable, and in technical circles, they were the most politically knowledgeable. But after five years in Washington, these emissaries between science and government were no longer regarded with awe in either camp. As internationalists, they early recognized that the atomic bomb would revolutionize world relations, and as sophisticated technologists they concluded that salvation lay in international cooperation and control, rather than in a hopeless attempt at monopoly or an all-out arms race. But though they were ahead of the politicians in pleading the case for freeing the atom from nationalist causes, they were considerably behind the internationalist designs of a significant segment of the rank and file that had produced the bomb. Large numbers of scientists at Chicago's Metallurgical Laboratory felt immensely alienated from the Washington leadership, initially on the issue of financial support once their plutonium work was completed. But this issue was dwarfed by the moral storm that raged in Chicago over the employment of the bomb. In 1945, with time on their hands, a committee of Chicago scientists, headed by James Franck, assessed the moral and political implications of the weapon they had helped build, and petitioned Washington against using it on a populated target. They argued that "a demonstration of the new weapon might best be made, before the eyes of representatives of all the United Nations, on the desert or a barren island. The best possible atmosphere for the achievement of an international agreement could be achieved if America could say to the world, 'You see what sort of a weapon we had but did not use. We are ready to renounce its use in the future if other nations join us in this renunciation and agree to the establishment of an efficient international control.' " [43]

On the other hand, in the capital of the nation that was leading the greatest war in history, the leaders of the bomb project showed no capacity for so disengaged a view. What was to be done with the weapon they had built in the course of an all-consuming four-year two-billion-dollar gamble? Should a harmless demonstration be attempted to warn the Japanese of the disaster they faced? Oppenheimer, Lawrence, Fermi, and Arthur Compton, meeting as the scientific panel of Henry L. Stimson's Interim Committee, weighed the issue, and concluded that in the interest of postwar international control and cooperation, the United States should inform its allies of the progress it had made on atomic weaponry. But they also concluded that the problems of arranging a harmless demonstration left no alternative to direct military use.[44] * Through two decades of ensuing debate, defenders of the decision have argued that in the turmoil of war and with the information then available, the decision was inevitable. That question itself is irrelevant here, but for developing policy mechanisms for employing powerful and little-understood technology, there is a lesson in the episode of the scientific advisory panel to the Interim Committee; namely, that the progenitors of a technology should not be the principal source of technical advice on the employment of that technology.

The Interim Committee accepted the advice of its scientific advisers, and the ensuing blasts over Hiroshima and Nagasaki instantly shattered the spirit of unity that had generally prevailed in the scientific community through the long war.

Within three months after Japan surrendered, some three thousand scientists, drawn mostly from Manhattan Project installations, had organized themselves into groups that eventually evolved into the Federation of American Scientists. They established representatives in lobbying posts in Washington, and commenced battle on an issue that came to be formulated in terms of civilian versus military control of the atom, internationalism versus monopoly, freedom versus secrecy. The

---

* The following account of these proceedings is provided in the official history of the AEC: "Oppenheimer could think of no demonstration sufficiently spectacular to convince the Japanese that further resistance was futile. . . . Conant suggested and Stimson agreed that the most desirable target would be a vital war plant employing a large number of workers and closely surrounded by workers' houses. Someone brought up the desirability of several strikes at once. Oppenheimer considered this feasible, but Groves doubted its wisdom." At midafternoon, with discussion completed on the prepared agenda, Stimson left, and Arthur Compton introduced the subject of the future of nuclear research at Chicago. (R. G. Hewlett and O. E. Anderson, *The New World, 1939/46,* Vol. 1, *A History of the United States Atomic Energy Commission* [Pennsylvania State University Press, 1962], p. 358.)

specific object of their efforts was the May-Johnson bill for postwar control of atomic energy, drafted by the War Department, and supported in its entirety or in large part by the wartime leaders. Never before had the scientific community been so actively involved in a public policy issue, so divided, or so public about its differences. "I must confess," Herbert L. Anderson, a Chicago physicist, wrote to a colleague at Los Alamos, "my confidence in our leaders Oppenheimer, Lawrence, Compton, and Fermi, all members of the Scientific Panel advising the Interim Committee and who enjoined us to have faith in them and not influence this legislation, is shaken. I believe that these worthy men were duped—that they never had a chance to see this bill. Let us beware of any breach of our rights as men and citizens." [45]

In late 1945, Groves quietly sought conciliation with the rebellious scientists, assuring them that he desired to continue research activities in the wartime laboratories, and that he was speeding work on declassifying reports on atomic research.[46] But right at the outset of Groves' peacemaking efforts, a new emotion-filled issue burst into the situation. On November 24, shortly before Groves was to meet with scientists at Oak Ridge and Chicago, *The New York Times* reported that American occupation forces in Japan had descended upon five Japanese cyclotrons, torn them apart with welding torches and explosives, and dumped them in the ocean. The cyclotrons, including a 220-ton machine copied after Lawrence's machine at Berkeley, were fundamental research instruments, employed in biological, medical, chemical, and metallurgical research, and had no military value. In a statement ringing with anguish, Yoshio Nishina, director of the Institute of Physical and Chemical Research, in Tokyo, reported that occupation headquarters had granted permission for operation of the cyclotrons upon recommendation of Karl Compton, who had visited Japan as head of a Scientific Intelligence Survey. "That not only this authorization would be rescinded but also that cyclotrons would be destroyed without adequate notice was almost beyond the limits of our credulity," Nishina declared.[47] *

* The cyclotron episode obviously demonstrates the obtuseness of at least those military minds immediately involved, but it also illuminates something else, namely, the ingenuity and perseverance of basic researchers in conducting basic research, even under the most difficult of circumstances. Nishina plaintively noted that the 220-ton cyclotron—which, it should be recalled, had no military value—had been completed in 1944. This, of course, was at a time when Tokyo, where it was located, was being reduced to rubble by massive B-29 raids, and

Now, on the eve of Groves' visit, the Association of Oak Ridge Scientists proclaimed, "Men who cannot distinguish between the usefulness of the research machine and the military importance of a 16-inch gun have no place in positions of authority." [48] Compton urged dismissal of the responsible officers; scientists at Harvard and MIT sent letters to the Secretary of War Patterson denouncing the destruction. Patterson later said he would take responsibility, while Groves explained that the machines were destroyed because of misinterpretation of an order from his office for the machines to be "secured." [49] With apparent sincerity, Groves, too, deplored the destruction, and took the responsibility upon himself, but the mood of the scientific community was such that nothing the general did could win him the favor of the rank-and-file scientists.

The cyclotron episode, coupled with revulsion against the atomic attacks on Japan, caused the rank-and-file scientists to view with suspicion any administrative designs of the wartime leadership, whether for the control of atomic energy or for the promotion of basic research. Harold Urey, one of the key scientists in the bomb project and later a recipient of the Nobel Prize, declared that the atomic energy bill initially favored by the administration was the "first totalitarian bill ever written by Congress. . . . You can call it a Communist bill or a Nazi bill, whichever you think is worse." [50] Two months later, Urey joined some two hundred other scientists in signing a statement that implied support of the administration's insistence upon political accountability of the Foundation. On the sensitive issue of the means of selecting the leadership of the Foundation, the statement simply said, "Although there is a serious division of opinion on the question of whether administrative responsibility should be given to a governing board or to a single administrator, it should be possible to devise a plan of organization which will meet the major objections to either alternative." But the signers then went on to say that their position was "in harmony" with the President's. [51] At the end of 1946, a White House staff memorandum stated, "The differences on National Science Foundation legislation be-

---

the entire Japanese nation was undergoing severe privation because of the American blockade. Amidst all these difficulties, however, Japanese physicists were busily working to construct a machine to study the composition of the nucleus of the atom. This work was not related to any effort to build an atomic bomb. Japan's brief effort in this field had been abandoned in 1941, in large part because of lack of uranium.

tween the Bush group and the Urey–[Harlow] Shapley–[Edward U.] Condon group are, very broadly speaking, the differences between a small 'inner' group closely allied with a few powerful institutions and large corporations (where most wartime research was conducted), and on the other hand, a larger group of scientists with interests spread throughout the nation and with a desire to avoid—insofar as possible— the concentration of research and the power to control it." [52] This analysis represented an attempt to fit the behavior of the scientific community into traditional ideological forms. The signers, in fact, included large numbers of scientists from institutions that were among the giants of wartime research, including, for example, Fermi, a member of the scientific panel of the Interim Committee.

Thus, the motives of the signers were mixed, and probably, in large part, rested on nothing more than a desire to put the Foundation into business on any sort of reasonable political terms.* But the effect of the statement was to reinforce the politicians' feelings that the scientific leadership had adopted an extreme position in its quest for federal support with extremely limited federal control.

Meanwhile, the very scientific leaders in whom the rank and file felt a loss of confidence were encountering a mixture of uncritical adoration and skepticism on Capitol Hill. On the one hand, Senator Russell of Georgia could ingenuously report, "My attitude toward scientists . . . is pretty much the attitude of the boy living in the country going to the country doctor. He thinks the doctor can do anything." [53] But Senator Connally, of Texas, apparently in reference to Conant, who was to serve as adviser to a high-level political delegation bound for Moscow, was heard to mutter "college professors" when he concluded that the administration had failed to keep the Senate properly informed of plans for international control of atomic energy.[54]

Louis Ridenour testified in behalf of the Federation of American Scientists for freedom of scientific information. He argued that there are no secrets in science, and in turn was lectured by Senator Hart, of Connecticut. "My prediction, Doctor, is that before you are as old as most of us here are you will have changed your views in that respect, and you will have many disagreeable surprises as to how easily others do learn our secrets." [55]

The battle over control of atomic energy raged on, absorbing the

---

* Reflecting on the episode two decades later, Urey said, "My views on NSF were, 'Let's get it going. The basis doesn't matter too much.' "

time and energy of those scientists with a disposition toward public affairs, and turning attention away from Bush's design for postwar support of basic research. In her definitive work on the scientists' role in the postwar political struggle over control of atomic energy, Alice Kimball Smith concludes: "In the closing weeks of the 79th Congress, the fight for civilian control of atomic energy so absorbed the more articulate members of the American scientific community that the tabling of science legislation caused no great outcry."[56] Late in 1946, under the auspices of the American Association for the Advancement of Science, nearly seventy scientific societies organized the Inter-Society Committee on Science Foundation Legislation. On paper, it was an impressive organization, ostensibly representing the interests of thousands of scientists. But, in fact, it was a device for effortless petitioning from a well-sheltered position, and its political impact was slight. Mrs. Smith concluded that "no group stood ready to put behind the passage of science legislation the concentrated effort" that the Federation of American Scientists had put into the atomic energy fight.[57]

In mid-1946 a Kilgore-Magnuson compromise on a science foundation bill, weighted toward the Kilgore concept, passed the Senate, but failed in the House. The following year, Senator H. Alexander Smith, of New Jersey, introduced a compromise, reflecting the Bush position. President Truman swiftly responded to its passage with a veto message declaring that the Smith bill "contains provisions which represent such a marked departure from sound principles for the administration of public affairs that I cannot give it my approval. It would, in effect, vest the determination of vital national policies, the expenditure of large public funds, and the administration of important governmental function in a group of individuals who would be essentially private citizens. The proposed National Science Foundation would be divorced from control by the people to an extent that implies a distinct lack of faith in the democratic process."[58] Don K. Price of Harvard, dean of American scholarship on science and government, who served in the Bureau of the Budget in 1945 and 1946, commented shortly after the veto message that the Smith bill "went as far as a Constitutional bill could go in creating an executive agency that would not in practice be effectively responsible to the President." Price also foresaw something that many of the statesmen of science, to their loss, were not to realize for more than a decade: that the Foundation "would have a great deal to gain in a positive way from the one thing that the Smith bill sought above all to prevent—a direct and responsible and intimate relationship with the Presidency." [59] It was not until 1950 that the Foundation came into existence—on Truman's terms. When officials of the newly established Foundation appeared before the House of Representatives to seek funds, they were told that, in the midst of the Korean War, neither manpower nor funds should be expended on something with so distant and uncertain a payoff as basic research. An appeal to the Senate brought a mere $225,000. That crushing pronouncement from the House fully vindicated the fears that had driven the leadership of the scientific community in their quest for a politically secure enclave. But, ironically, a great incongruity accompanied the ideological defeat. For in the years following the war relatively great prosperity came to the American scientific community. Through complex motivations, the military—particularly the Navy, with its long scientific and technical tradition—became the subsidizer of American basic research. And the military assumed this role largely on its own initiative, for the leadership and unity of purpose that had bound American sciences throughout the war no longer existed. The American scientific community had escaped a return to its prewar poverty, but it was also rudderless in its affluence.

NOTES

1. "Some Problem Areas in the Relationships Between Government and the Universities," address to National Academy of Sciences, October 13, 1964, *News Report,* National Academy of Sciences–National Research Council, November-December 1964, p. 90.
2. *Atomic Quest, A Personal Narrative* (New York: Oxford University Press, 1956), p. 72.
3. Vannevar Bush, *Science, The Endless Frontier* (1945; reprinted by National Science Foundation, 1960), Appendix 3, pp. 128–29.
4. James P. Baxter 3d, *Scientists Against Time* (Boston: Little, Brown and Co., 1946), Appendix C, p. 456.
5. Compton, *Atomic Quest,* p. 196.
6. *Ibid.,* p. 80.
7. *Ibid.,* p. 81.
8. R. G. Hewlett and O. E. Anderson, *The New World, 1939/46,* Vol. 1, *A History of the United States Atomic Energy Commission* (University Park, Pa.: Pennsylvania State University Press, 1962), p. 324.
9. Correspondence, Nichols to Groves, December 15, 1944, Manhattan Engineering District collection, National Archives.
10. *New York Times,* February 19, 1956.
11. *Legislative Proposals for the Promotion of Science,* Texts of Five Bills and

Excerpts from Reports, Subcommittee on War Mobilization, Senate Committee on Military Affairs, August 1945, "Summary Comparison."

12. *Hearings on Science Legislation*, Subcommittee of the Senate Committee on Military Affairs, 1945, p. 111.
13. *Ibid.*, p. 203.
14. *Surplus Material Research and Development, Hearings*, House Select Committee on Postwar Military Policy, 1945, p. 129.
15. Compton, p. 57.
16. Bush, p. 3.
17. *Atomic Energy, Hearings*, Special Committee on Atomic Energy, United States Senate, 1945, p. 46.
18. Bush, p. 34.
19. *Ibid.*, p. 33.
20. *Ibid.*, p. 40.
21. *Ibid.*, pp. 142–43.
22. *Ibid.*, p. 23.
23. *National Science Foundation, Hearings*, House Committee on Interstate and Foreign Commerce, 1946, p. 12.
24. "The Scientists: A Group Portrait," *Fortune*, October 1948, p. 176.
25. *Hearings On Science Legislation*, 1945, p. 509.
26. *Ibid.*, p. 785.
27. Bush, p. 61.
28. *Ibid.*, p. 57.
29. *Ibid.*, p. 63.
30. *Ibid.*, p. 16.
31. *Ibid.*, Appendix 4, p. 160.
32. *Ibid.*, p. 8.
33. *Ibid.*, p. 20.
34. *Hearings on Science Legislation*, 1945, p. 331.
35. *Ibid.*, p. 332.
36. *Ibid.*, p. 628.
37. *Ibid.*, p. 982.
38. *Ibid.*, p. 471.
39. *Ibid.*, p. 342.
40. *Ibid.*, p. 623.
41. *Ibid.*, pp. 332–33.
42. *Legislative Proposals for the Promotion of Science*, August 1945.
43. The Franck Report, reprinted in Alice Kimball Smith, *A Peril and a Hope* (Chicago: University of Chicago Press, 1965), pp. 566–67.
44. Hewlett and Anderson, p. 367.
45. Quoted in Alice Kimball Smith, p. 140.
46. Smith, pp. 349–52.
47. Yoshio Nishina, "A Japanese Scientist Describes the Destruction of His Cyclotrons," *Bulletin of the Atomic Scientists*, June 1947, p. 145.
48. Smith, p. 354.
49. Leslie R. Groves, *Now It Can Be Told* (New York: Harper and Brothers, 1962), pp. 367–72.
50. *New York Times*, October 31, 1945.
51. *Science*, January 4, 1946, p. 11.
52. Harry S Truman Library, 192-E.
53. *Atomic Energy, Hearings*, 1945, p. 170.
54. Hewlett and Anderson, p. 473.

55. *Atomic Energy, Hearings,* 1945, p. 547.
56. Smith, p. 444.
57. Smith, p. 521.
58. Reprinted in *National Science Foundation, Hearings,* House Committee on Interstate and Foreign Commerce, June 1948, p. 24.
59. Don K. Price, "The Deficiencies of the National Science Foundation Bill," *Bulletin of the Atomic Scientists,* October 1947.

VII
The Reluctant Leaders

Unfortunately, the scientific leaders in the [atomic bomb] project who would normally have been the spokesmen for their colleagues were preoccupied with getting back to their peacetime occupations. . . ."

GEN. LESLIE R. GROVES, in *Now It Can Be Told,* 1962

I think it is a marvelous thing that the government gets as much [advisory] service as it does from the scientific community. Frankly, if I did private consulting with the time that I devote to government committees, it would certainly net me more than my regular salary.

A Nobel Laureate in Physics, 1965

For the scientific community, the decade following World War II was a time of organizational genesis. These were the years in which there came into being virtually all of the institutional arrangements and economic patterns that prevail to this day in relations between science and government. It was a period in which the once-monastic community of science was simultaneously wooed, fattened, humiliated, elevated to honor and influence, and ignored. It was the time of the Oppenheimer case, the birth of the National Science Foundation, the beginnings of a science advisory system for the White House, and the entrenchment of the scientific community as a permanent—and ever costlier—ward of government. As we look back on the first postwar decade of relations between science and government, it is evident that something between seduction and rape repeatedly occurred, but at various points it is by no means certain which party was the aggressor and which the victim. What is certain, however, is that science needed the government treasury and government needed the talent and works of science. Respectively following the delicate traditions of science and the sometimes oxen ways of government, the parties collaborated in a wondrously complicated and ever-growing relationship, so amorphous in its outlines, and so unlike anything else in public affairs, that as late as 1963 a perspicacious

125

political writer concluded: "Nobody has—or could—come up with a readily comprehensible table of organization to explain the labyrinth of [scientific] agencies, foundations, consultantships, academies, and committees that has grown up in Washington in recent years." [1]

The irony of it was that, having demonstrated during World War II a brilliant capacity for mobilizing itself and serving the nation, the scientific community showed very little capacity, in the decade following the war, for determining its own fate. On the other hand, there was scarcely any incentive to develop this capacity, for relative affluence came easily to postwar science, and if the conditions accompanying it did not fully dovetail with the community's ancient ideology of independence, they at least were not so bad that the divergence could not be borne.

Perhaps the most striking feature of the first postwar decade was that military services, recalling their prewar neglect of science and technology, and squabbling among themselves for jurisdiction over atomic-age weaponry, were now eager for an intimate relationship with the men and works of science and technology. Their embrace of the prewar orphan was of such intensity that in 1952 Conant observed that "what I am concerned with is not the technological conservatism of the men in uniform in 1940, but the almost fanatic enthusiasm for research and development of their successors in 1952. . . . The Defense Department, in regard to research, is not unlike the man who sprang onto his horse and rode off madly in all directions." [2]

But it was not simply a case of the military wooing and the scientists responding. If the military was determined not to repeat its error of prewar neglect, there were many scientists who, for a variety of reasons, were wholly willing, often eager, to make their skills available to government—at least just short of the point of becoming full-time government employees. A sense of public responsibility, guilt about the bomb, the excitement of being counsel to the upper echelons of government—all these served to create that postwar Washington phenomenon, the scientist-commuter. But a foremost factor in the match of science and the military was simply that many scientists, having done their bit in World War II, quite easily made the transition to the Cold War. Some of their colleagues, centered in the group that helped fill the pages of the *Bulletin of the Atomic Scientists,* the conscience journal of science, would have no part of it. But when the military sought scientists for research or advice, there was no dearth of willing talent. To the contrary, the Cold War in combination with recollections of how well science had performed during World War II, inspired many scientists to thrust themselves into military policy making. The concerns of these volunteers appeared to cover a wide spectrum. At one extreme was the political right wing of nuclear science, based at the Radiation Laboratory at Berkeley, and led by Edward J. Teller, Ernest O. Lawrence, and his two close associates, Luis Alvarez and Wendell M. Latimer—all of whom shared in one domineering conviction: the Soviets were a mortal threat to the nation, and supremacy in the arms race was the only answer to the threat. In close alliance with the Air Force, which was seeking to remain the principal custodian of the nation's nuclear weaponry, they advocated the swift development of hydrogen weapons as the only safe response to the Soviets' development of conventional nuclear weapons.* [3] At the other extreme of this spectrum of weapons scientists was a Cambridge-centered group, including Jerome Wiesner and his MIT colleague, Jerrold R. Zacharias, a fellow Rad Lab alumnus, that was urging the expansion of the nation's conventional military forces to avoid exclusive dependence upon nuclear weapons in the event of confrontation with the Russians.[4]

The effect of this conflict was simply a further inflammation of the arms race, for both camps won, with the result that strategic doctrine embraced hydrogen weapons as well as an expansion of limited-war forces. But, for the purpose of examining the politics of science during the first postwar decade, the point to be noted here is that, amidst these conflicts over strategic policy—and, in fact, because of them—an intimate relationship developed and thrived between the military services and the nation's most accomplished and influential practitioners of basic research. But, again for our purposes, it is also important to note that the scientists were looking outward, away from their professional community—which was prospering, without very much effort, on money

* It was Latimer who led the unprecedented floor rebellion in 1950 that resulted in Conant's defeat and Detlev Bronk's election to the presidency of the National Academy of Sciences (see Chapter I, p. 14). It is probable that the main motivation for this remarkable episode in the traditionally decorous Academy was the realization that Conant, though one of the most prestigious figures of science, had taken little part in Academy affairs and was heavily engaged with his duties as President of Harvard and frequent adviser to the government. But it should be observed that while Latimer and his colleagues advocated a wide open arms race, Conant had been campaigning for arms control and international supervision of atomic energy as the surest path to national security.

that the military provided for research in the universities—and toward the far more compelling matters of national military strategy. Thus, in 1948, MIT, the AEC, and the Navy collaborated in the first of a little-known series of assemblages that were to bring the scientists still deeper into military affairs. Known as "summer studies," these were meetings of outstanding scientists, gathered together at a relatively quiet place, to give their undivided attention to specific tactical and strategic problems. (Two Navy scientists, long involved with summer studies have referred to them as "a kind of latter-day gathering of minutemen, called out to wage a brisk season's campaign on one of this country's tougher problems.") [5] The first of these, Project Lexington, held in the township of Lexington, Massachusetts, was concentrated on atomic-powered flight and its military implications. Next came Project Metcalf, in 1951, under a Navy contract with Harvard. The purpose of Metcalf was to study infrared detection. The leader of this study was Donald F. Hornig, who became presidential science adviser in 1964. Other summer studies included Project Charles, conducted by MIT for the Air Force, to study continental air defense and Project Vista, conducted by Caltech under contract to the Army to study the tactical and strategic implication of battlefield use of atomic weapons. Still others dealt with underseas warfare, civil defense, and a great variety of military-technological problems. (Jerrold Zacharias, of MIT, a key figure in many of these studies, originated the quip, "Summer Studies and some are not.")

It cannot be said that the scientists and the military were wholly compatible with each other. In fact, it was Oppenheimer's reluctance to pursue the development of the hydrogen bomb that inspired the Air Force to set in motion the events that led to his exclusion from strategic deliberations. What we have to note, however, is that while conflict prevailed outside the basic research community, and while inhabitants of the community regularly sallied forth to take part in that conflict, the community itself managed to remain a relatively sheltered enclave as far as its own substantive proceedings were concerned. While Lawrence and Alvarez battled in behalf of the hydrogen bomb, they were also occupied with building accelerators; while Rabi struggled in behalf of a diversified military force, he was building Brookhaven as a great center for basic research. The remarkable fact is that through all these struggles, and others that followed, the pure science community remained more or less immune in its internal functioning. While the Air Force battled against the arms control scientists centered in Cambridge, it was simultaneously increasing its expenditures for academic basic research, a good deal of which went to MIT. Despite the most ardent budget-cutting intentions of the Eisenhower administration, federal expenditures for academic basic research constantly rose, and, on occasion, with the connivance of the Air Force itself. Thus, the official history of the Air Force Office of Scientific Research (OSR) relates that in 1954, the Bureau of the Budget decreed that the Air Force had neither the right nor the need to support academic basic research. In response, the history relates, OSR's "budget was moved under a line item for the B-58. And, for all the Budget Bureau knew, the $4.7 million it approved was for research connected with the development of this aircraft, clearly within the realm of applied research. But, in reality, this money was handed over to OSR to use, as originally planned, for basic research." [6]

There was turbulence within this enclave, but it was turbulence resulting from forced-draft growth financed with government money. There was despair and anguish over the way Oppenheimer and others were treated in the public arena outside the scientific community, and, in a few instances, the imposition of loyalty and security regulations made it impossible for persons to continue with their work. There is no question that there was fear and ugliness, but as far as the actual conduct of scientific research was concerned, there was never anything resembling, for example, the Stalin-Lysenko domination, and destruction, of Soviet biology.*

In 1953, Mrs. Oveta Culp Hobby, the newly installed Secretary of Health, Education, and Welfare, instituted loyalty regulations that led a number of scientists, among them Linus Pauling, the Nobel laureate chemist, to part with the research grants they held from the Public Health Service. However, the problem of enabling them to continue their work was easily solved. The newly established National Science

---

* One of the most loathsome cases of witchhunt intrusions into the scientific community involved the Chinese missile pioneer, Tsien, over whom the McCarthy madness swept in the 1950s. Though his security clearance was lifted and FBI and Immigration Service agents hounded him, an important fact to note, I believe, is that Caltech, where he served on the faculty, went to extraordinary lengths to provide a haven for him and fight his case. If Tsien had chosen, he could have remained there and continued his productive career; the work he was interested in was not in classified areas. He chose, however, to return to China, where he was warmly welcomed and put in charge of developing missiles. For an excellent account of this episode see Milton Viorst, "The Bitter Tea of Dr. Tsien," *Esquire,* September 1967.

Foundation quietly came into the picture and provided them with financial support for the continuation of their research.*

During that same year, the Director of the National Bureau of Standards stood by his insistence that a top-selling battery additive, AD-X2, was worthless. Laboratory tests at the Bureau had demonstrated that the product did not live up to its claims of rejuvenating or in any other way improving batteries. The manufacturer, facing loss of postal privileges, sought relief by appealing to Congress and to the business-minded Eisenhower administration. Secretary of Commerce Sinclair Weeks, whose department had jurisdiction over the Bureau, decreed that good science was not necessarily good business, and demanded a new decision or the resignation of the Bureau's director. The ensuing public outcry was so great that Weeks swiftly retreated, reinstated the director, and thereafter left science to the scientists.

Eisenhower came to office pledged to reduce government expenditures. But medical research, which had been slow to get in on the growth of federal money for science, had at last found its way to political influence. A close collaboration had been formed by the mid-1950's between voluntary health associations, such as the American Cancer Society and the American Heart Association, and two health-minded congressmen, Representative John Fogarty (D-Rhode Island) and Senator Lister Hill (D-Alabama). Each headed his chamber's appropriations subcommittee for the Public Health Service, and each was determined to pour a fortune into medical research, regardless of the wishes of the Eisenhower administration and the fears of many of the conservatives who ruled the nation's medical schools—whereupon the budget for the National Institutes of Health, the research arm of the PHS, began to accelerate at an incredible pace. In 1955, NIH received $81 million; in 1957, the sum was $213 million; in 1959, the appropriation rose to $324 million; when Eisenhower left office it was near the three-quarters-of-a-billion mark.

Throughout American society, the affairs of pure science were deemed to be solely the concern of the men of pure science, except to the extent that the public generally felt that more was better. All that was required of government was that it provide the money, and this it

---

* A distinguished biochemist, who was later to receive the Nobel Prize, recalls that he and a number of colleagues considered turning in their government grants to protest Mrs. Hobby's actions. "But then we realized," he said, "that if we did that we might not be able to continue our research."

did, in ever-increasing amounts; for, on the basis of the World War II experience, there was no doubt that, when properly fed, science and technology paid off. The role of basic research in this payoff was a somewhat murky matter, for the men of science themselves could do no better than argue that their work was valuable but unpredictable. And, at times, they had to argue hard before uncomprehending congressional committees. But the military had more or less of a blank check on Congress; the AEC was always generously treated. Science was involved in a grand intimacy with both, and if funds for basic research were not forthcoming through the purely civilian National Science Foundation, they could always be squeezed out of the huge amounts appropriated for weaponry.

Relative to the amounts spent on developing hardware, the sums for pure science were piddling, but in terms of prewar budgets they were immense. According to one estimate, in 1954 the federal government spent $121 million for basic research, mostly in universities or in federal research centers operated by universities. Of this amount, nearly $80 million came from agencies with responsibilities that were exclusively or primarily military: the Atomic Energy Commission (whose principal focus at the time was a rapidly expanding weapons program), $40.4 million; the Department of Defense, $25.5 million, and the National Advisory Committee for Aeronautics, $13.7 million.[7] In some scientific fields, federal funds comprised the great bulk of the total available support. For example, in the academic year 1953–54, it was estimated, the nation's universities spent $56 million for research in the physical sciences. Of this amount, $47 million was provided by federal sources,[8] and most of this, it may be presumed, came from the three above-mentioned agencies.

Since the scientific community had so long espoused an ideology of independence, how did the military services so easily occupy a dominant position in the economy of civilian science? The answer is not simply that they bought their way in—though the record shows that few quibbles were raised when they came bearing money. Rather, the genesis of military involvement in the support of science was the consequence of a number of elements that easily dovetailed. The experience of war had transformed the military services into zealous believers in the power of science and technology. But they were infected not only by science's power, but also by science's ideology of independence. In

1946, Army Chief of Staff Dwight D. Eisenhower issued a directive that proclaimed: "Scientists and businessmen contributed techniques and weapons which enabled us to outwit and overwhelm the enemy. . . . Scientists and industrialists must be given the greatest possible freedom to carry out their research. . . . Scientists and industrialists are more likely to make new and unsuspected contributions to the development of the Army if detailed instructions are held to a minimum." [9] * Thus, the military services displayed the zeal of the converted true believer; they would support science, and they would support it more or less on its own terms. The effect of this decision was simply to remove any economic compulsion for the scientific community to develop solidarity and campaign for government support on its own terms. There was no need to fight when the prize was simply handed to it.

The Manhattan Project physicists, who were the most cohesive and politically active of all the scientists, produced superb rhetoric on the necessity for scientific independence. And they delighted in tales deprecating General Groves' intellectual qualities. Apparently he sensed their presumption of superiority, for he later wrote, "Experience had shown that the scientists were most critical of anyone whose mental alertness did not equal or excel theirs." [10] But Groves was pouring leftover wartime money into research. Four months after the surrender of Japan, he agreed to provide up to $170,000 to help continue the construction of the 184-inch cyclotron that Ernest O. Lawrence had set aside at the outbreak of the war.[11] Looking back on the early postwar period at the Berkeley Radiation Laboratory, Luis Alvarez commented, "We ran it with a big barrel of greenbacks." While rhetoric flowed on how government should support science, Groves was supporting it with sums that were staggering by prewar standards. At the end of the war, he established a seven-member Advisory Committee on Research and Development, consisting of Robert F. Bacher, Arthur H. Compton, Warren K. Lewis, John R. Ruhoff, Charles A. Thomas, Richard C. Tolman, and John A. Wheeler—all important figures in the wartime program, all eager to get on with a great postwar research program. The committee recommended expenditures of $20 million to $40 million for the fiscal year starting 1946, depending upon the availability of

---

* It is interesting to recall that two decades later, when the Defense Department made a retrospective study of the sources of new weapons technology, it came to the conclusion that "the contributions from recent post-World War II undirected research in science was very small." See Chapter II, p. 31.

manpower. Groves responded by scheduling expenditures of over $72 million.[12] Most of the sum was to keep the weapons program going, but a lot of it was for fundamental research.

Rabi, after having spent the entire war at MIT's Radiation Laboratory, returned to Columbia to find that its long-standing eminence in physics was shattered. Compton had succeeded in inducing Fermi to take a peacetime post at Chicago. Groves was providing money for Chicago's ambitions, while, at Berkeley, the pioneer center for particle physics, Lawrence was planning and building new machines, also under subsidy from Groves. Columbia lacked both the space and the aggressive administration required for these great implements of postwar physical research. But Rabi had the prestige of a Nobel Prize, initiative, and the respect of his colleagues. Groves advised him that while the lame-duck Manhattan Project still had funds, he should move quickly, harmonize the inchoate postwar aspirations of the major East Coast institutions, and submit a proposal for a cooperative venture. Rabi swiftly organized nine eastern universities * into a consortium called Associated Universities, Incorporated, and began a search for a site for a large-scale cooperative laboratory for nuclear research. Meanwhile, he was also seeking to recruit talent for the eastern renaissance, and his travels even took him to Oak Ridge, Tennessee, ironically, one of the Manhattan Project centers which Groves, the underwriter of Rabi's aspirations, was desperately trying to hold together. "I went down to Oak Ridge," Rabi recalls, "and made a speech. 'Boys, the war is over. Why should you be stuck here? Why don't you come to an intellectual center?' It was received enthusiastically."

The site for the cooperative laboratory quickly became a subject of controversy. A proposal to locate it at Cambridge was opposed by a Cornell representative on the grounds that "MIT's already gotten enough out of the war." Groves responded to the controversy by warning that if the universities did not get together, they would get nothing. A site selection committee then fixed upon a surplus military base on Long Island, Camp Upton, where the availability of Army buildings provided a head start for what was to become the Brookhaven National Laboratory, site of what eventually was to be the most powerful particle accelerator in the country.

In large and small amounts, wartime funds trickled into the univer-

* Columbia, Cornell, Harvard, Johns Hopkins, MIT, Princeton, University of Pennsylvania, Rochester, and Yale.

sities, thus helping to cushion the transition from war to peace. For example, in 1945, OSRD, in the process of winding up its affairs, transferred forty-four contracts to the National Institutes of Health to support the continuation of university-based medical research.[13]

Meanwhile, a group of young naval officers, nicknamed the "Bird Dogs," who had been deep in the administration of the Navy's wartime research effort, took encouragement from the Navy's traditionally friendly disposition toward science and advanced technology. They encountered virtually no opposition when they proposed that the Navy create what in effect would be a national science foundation, to be known as the Office of Naval Research (ONR). Its object would be to support basic research, principally in the universities, not because the Navy would immediately benefit, but because over the long run, they reasoned, the Navy would be strengthened by the general advancement of science. Uniformed officers headed the organization, but its policies and programs were formulated and carried out by civilian scientists to whom the Navy gave free rein. ONR's birth was made relatively pain-free by a fact of politics that remains in force to this day: place a military label on a scientific venture, and even if the proposal has only remote, or possibly no military significance, political barriers dissolve. In 1948 with the NSF proposal vetoed and still languishing, the Navy could report without arousing any noticeable political interest: "Most of the work [of ONR] is in our universities and nonprofit laboratories. The programs cover the physical, medical, and biological sciences. Today we have approximately 700 research projects in over 150 universities and nonprofit laboratories. About 2000 scientists and over 2400 graduate students are involved. The annual expenditure for these programs has reached a level of over $18 million and is estimated to reach a $20-million level by the first of the coming year." [14]

The Navy's swift movement into basic research was a magnificent windfall for American science, for the initiative came wholly from within the Navy, though once it was decided to establish ONR, the wartime leaders of science were brought into close collaboration, and one of OSRD's top administrators, Alan T. Waterman, was appointed Chief Scientist of ONR. Nevertheless, with admiration, Bush said, "They [the Navy] did it themselves." And Rabi states, "The Navy saved the bacon for American science." But perhaps even more remarkable than the creation of ONR was the extent to which the scientists' ideology of independence had permeated the Navy, to the point where, at the outset, the Navy was not certain whether the academic scientific community would welcome military support. In 1945, when the nascent ONR was evolving from the Navy's wartime research organizations, the scientists' litany of freedom aroused deep insecurities in the aspiring new patron of science. "There remained one serious hurdle, outside of Congressional action [which proved to be no hurdle], before the establishment of ONR could become meaningful," the founders of ONR wrote years later. "This was to get the universities, where the majority of basic research is performed, to be willing to accept Navy contracts. In this struggle Capt. [R. D.] Conrad [director of ONR's planning division and one of the principal architects of the office] became the recognized leader. Accompanied by various 'bird dogs' Conrad visited many top universities in the winter of 1945. There was a definite feeling on the part of the scientists after four years of war to wish to forget the Navy and return to former pursuits. But Conrad was able to crush all opposition by making superb speeches around the country, and by working with legal and contract people to pioneer an acceptable contract system. This would permit one overall contract with a university with new task orders to be attached as agreed upon, permit basic research to be contracted for, and permit the work to be unclassified and published. Once the legal eagles got this worked out, there was no holding the persuasive Conrad, and he was quickly able to get such institutions as Harvard, Chicago, University of California, California Institute of Technology, and MIT to agree to Navy work." [15]

Some scientists were, in fact, reluctant to accept military money. George B. Kistiakowsky, of Harvard, for example, recalls that he hesitated "because I was worried about real science being guided into militaristic channels by military support. I made my application only under the assurance that I could do any research I wanted to do within the broad area of my interest." And he added that "ONR never dictated anything to me about my research. There has been no more control there than there has ever been from NSF, or for that matter, from the Rockefeller Foundation." Three years after the end of the war, *Fortune* described ONR as the "smartest and most carefully organized of the new military agencies . . . without this support, basic research in the U.S. would be at its lowest ebb in history. . . ." [16]

Ideologically, the scientists were addicted to Bush's concept of the self-government of science, and a few scientists, such as Kistiakowsky, approached the newly established military patron with caution. The scientific community as a whole, however, was not inclined to follow principle out the window. Rhetoric flowed, but the reality was that the military services and the AEC found willing and grateful recipients as fast as they could parcel out funds. In a printed exchange in 1947 titled "Military Support of American Science, A Danger?" [17] Louis N. Ridenour, dean of the graduate school at the University of Illinois, and, later, chief scientist of the Air Force, pointed out that: "The decline of the dollar notwithstanding, science has more money now than it had before the war; and the increase is wholly due to the backing that the Armed Forces have seen fit to give." Did military support constitute a danger to science? Ridenour advised his colleagues to be alert, but concluded that if the military funds were provided free and clear, "I can see no reason for refusing the . . . help—granted it is forthcoming—in doing the scientific work that one would have tried to accomplish even without such help." Replied Albert Einstein: "Is it at all reasonable that the distribution of funds raised for these [scientific] purposes from the taxpayer should be entrusted to the military? To this question," Einstein asserted, "every prudent person will certainly answer: 'No!' For it is evident that the difficult task of the most beneficent distribution should be placed in the hands of people whose training and life's work give proof that they know something about science and scholarship." Philip Morrison, a Cornell physicist, warned, "We cannot tie science to the military and hope to see it used for peace, no matter how ingeniously we write the contracts. . . ." Norbert Wiener, whose contributions to computer theory and design were, ironically, to have the most far-reaching military applications, flatly declared: "The Armed Services are not fit almoners for education and science. They are run by men whose chief purpose in life is war, and to whom the absence of war, even though a war is almost certain to engulf them personally, is a frustration and a denial of the purpose of their existence. . . . What makes a man a good soldier generally makes him a bad scientist, and a totally unsuitable administrator of science."

Such was the rhetoric, much of it tinged with the scientists' ancient ideology of independence and recollections of indignities, real or imagined, suffered at the hands of the military during the war. But the relationship of rhetoric to reality was another matter. What were the motives of the military? Undoubtedly, better weaponry, plus good relations with the newly arrived men of influence in strategic affairs, the scientists. What were the motives of the scientists? The answer to this question is far more complicated. A substantial segment of the community was simply gripped by the profession's traditional zeal for basic research, and was unconcerned about the source of support so long as the money came without restrictions. Could the military be counted on to provide, and to continue to provide, money on this basis? The available evidence was that, by and large, it could. But always present was the scientists' realization that their profession had a well-developed capacity for eluding the undesirable intentions of its patrons. Ridenour succinctly stated the case in an anecdote: "I have a friend who is a band spectroscopist on the faculty of a large state university. He has been particularly interested in the band spectrum of the element nitrogen. He once said to me: 'When the representatives of the state legislature visit me, I always tell them I am trying to make better fertilizer.' There is to be sure," Ridenour continued, "nitrogen in fertilizer and knowledge is power. It is just conceivable that my friend's investigations of the band spectrum of nitrogen may some day affect the fertilizer industry in some unexpected way. But it is undeniable that his interest is in spectroscopy, in and of itself." [18]

There was another element, however, in the scientists' disposition for intimacy with the Armed Services and the military-related federal agencies, such as the AEC. And that had to do with their own fervor for American technological supremacy, particularly in Cold War matters, and their belief that basic research was an essential component of this quest for supremacy. Thus, Willard Libby, the Nobel laureate chemist who served as an AEC commissioner during the Eisenhower administration, and, in the 1964 presidential campaign as an adviser to Barry Goldwater, offers the following explanation for federal agencies seeking close ties with "top scientists." States Libby: "You see, here's what you do: it's a kind of modern piracy. When I was on the AEC I carefully 'ground in' all the top talent I could find in the physical sciences and we had near 99 percent of them. You see, you know who they are through your connections and you can get them, and you just hold them. And you give them everything they need and they get loyal to you and pretty soon you've got them, you see. This is what can be done, and this is the great battle among the agencies."

In any case, in the first decade after the war the irony of it all was astonishing. Bush and his colleagues had proposed the establishment of NSF to provide academic basic science with its own specially insulated access to the federal treasury. The political community had balked, and the proposal was becalmed. Meanwhile, the Navy had become, on its own initiative, a major subsidizer of science, and on terms that conceded all to the scientists' traditional insistence upon freedom and independence. As for the AEC—this was the only existing postwar federal agency in whose genesis the basic science community had played a major role. Abandoning the laboratory for an unprecedented foray into political activism, the scientists had supposedly won a great victory, pushing through legislation that provided for civilian, rather than military, control of the atom. Nevertheless, while the Navy, because it was so inclined, yielded all to the ideology of science, the AEC, despite the scientists' role in its genesis, granted all in the way of money, but little in the way of ideology. As Walter Gellhorn points out in a study of the impact of internal security programs on civil liberties, "In the twelve months between November 1947 and November 1948, 1,936 research reports were produced in the laboratories which the AEC controls. Of these reports, over three-quarters (1,567) were deemed by the Commission to contain information that must be kept in a restricted category, and accordingly the reports have been concealed from all but a few selected persons. Two hundred and ten of the research reports related to health and biology; in this group 176 papers, 84 percent of the total, were 'classified' and held to be nonpublishable. This is especially interesting because research in the fields of medicine and health have traditionally been 'open.' Even during the years of active war, the military authorities agreed that publication of new medical findings should be encouraged. . . ." [19] In contrast, virtually all research supported by the Office of Naval Research was free to be published in the open scientific literature.

OSRD was a masterpiece of political planning and maneuver. But in the aftermath of this triumph, chance, rather than design, prevailed in the political affairs of the scientific community. Why?

The largest part of the answer is that the scientific community, though now well en route to becoming a financial ward of the federal government, did not conceive of itself as an interest that required representation before the federal government. Who spoke for science in the postwar decade? No one. During World War II, Bush and Conant, and perhaps a few others, were properly regarded as the scientific community's emissaries to Washington. Their task, as conceived by themselves and accepted in practice by Roosevelt, was twofold: to identify those areas in which science and technology might contribute to the war effort, and to enhance those conditions which contributed to the productivity of the nation's researchers. Thus, the creation of the politically insulated OSRD, the battles with General Groves over what the scientists considered stifling security measures in the Manhattan Project, and the incessant struggles over draft deferments for researchers. Organizational structure, discipline, and chain of command were, and are, normally incompatible with the anarchic, rambling ways of science. In the crisis of war, however, these fetters could be accepted, in fact, self-imposed, by the nation's research enterprise, but with incredible rapidity the war-born unity fell apart almost immediately at the end of the conflict.

The desire of the medical researchers to be excluded from Bush's all-encompassing Foundation was perhaps the most vivid symbol of the political fragmentation of American science. But other factors contributed even more to creating a vacuum of scientific leadership in Washington. When peace returned, the political leadership more or less recognized the value of science and technology for national strength, but this recognition was not accompanied by any disposition to bring full-time scientific talent in to the upper policy-making councils. It was not until 1957, following the crisis of Sputnik, that the White House was to be served by a full-time science adviser, for, in the absence of crisis, there was no demand from the political side for science to maintain a vigorous, continuing presence in Washington for the purpose of representing the scientific community. But, of perhaps greater significance, there was virtually no initiative on the part of the scientific leadership to establish a clearly identified place for science in the policy-making councils. Though government was now utilizing and subsidizing science and technology to an extent that was inconceivable in the prewar days, the leaders of science manifested no desire to become part of the political process that had suddenly brought affluence to their recently impecunious profession. Bush, it will be recalled, had proposed the creation of a Science Advisory Board that would give academic scientists influence over the conduct of research by government bureaus, but outside of this characteristic gambit for assuring academic dominance, his grand design for the postwar collaboration of science and government made no provision for bringing scientists to the high policy councils as representatives of science.

From the perspective of the present, when full-time scientific expertise is firmly established as part of the White House staff apparatus, Bush's treatment of the issue appears curious, but in terms of the traditions of science, the scientific community's vision of itself, and the state of the collaboration between science and government, it was not out of order in that period. Government saw no need for a continuing scientific presence in Washington, but even more important, as the war came to an end, those who had led science throughout the conflict were eager to return to their own affairs.

As early as the spring of 1944, Bush had warned, "The OSRD is a temporary war organization which automatically goes out of existence at the end of this war. . . ." [20] For speeding the enactment of his NSF designs, there was some tactical merit in a rapid dissolution of OSRD. But though OSRD lingered on until the end of 1947, it was in lame-duck status without influence or new funds. Meanwhile, the NSF proposal was politically becalmed. Bush himself kept one foot in government through chairmanship of the Pentagon's part-time Joint Research and Development Board, whose creation he had recommended as a means of assuring technological rationality in the defense establishment. But part-time service was not equal to the task of adjudicating the interservice strife that began to rage over assignment of postwar weaponry. In 1949 Bush resigned, and thereafter had little to do with the affairs of science and government.* Conant and Karl Compton both gave generously of their time for advisory services to government, but their principal concerns were the postwar building of their universities. Following his defeat in 1950 for the presidency of the National Academy of Sciences, Conant began to disengage himself from science policy affairs. He served briefly, and apparently without much enthusiasm, as the first chairman of NSF's chief advisory board when the Foundation

---

* Bush's reputation as a military prophet did not long survive the war in which he had performed so brilliantly. In 1949, he published *Modern Arms and Free Men* (New York: Simon & Schuster, Inc.), a work which alternated between deprecating the military for its prewar neglect of science and its postwar addiction to it. But the military's weaponeers were making far greater progress than Bush, now an outsider, realized. "The atom bomb," he wrote, "cannot be subdivided. This is inherent in the physics of the situation. . . . There will be no shells from guns carrying atomic explosives, nor will they be carried by marine torpedoes or small rockets. . . ." (P. 106.) Bush also discounted the likelihood of intercontinental rockets, writing, "There will be no such thing for a long time to come. . . . By the time there are intercontinental missiles, if ever, we shall have an entirely new set of circumstances to consider." (P. 116.)

came into existence in 1951. In 1953, however, he left Harvard to become United States High Commissioner in West Germany, and he was never again concerned with the problems of science and government.[21] Jewett, the fourth member of the group that had brought science to the wartime service of government, was sixty-eight years of age at the end of the war. He broke with his colleagues on the issue of government support of science. Taking an elitist view, he argued that a large infusion of funds would dilute the quality of American science. Philanthropy and industry, he said, could meet the need: ". . . The amounts of money involved in support of first-class fundamental science research —the only kind that is worthwhile—are infinitesimally small in comparison with national income. The values of this kind of research are measured in terms of the minds of men, not in quantities of money." [22] But Jewett was a minority of one.

Oppenheimer left Los Alamos for Caltech. DuBridge left the Radiation Laboratory, briefly returned to the University of Rochester, and then became president of Caltech. When the Atomic Energy Commission came into existence, many of the wartime scientific leaders—among them Oppenheimer, Conant, DuBridge, Fermi, Rabi—served on its highest scientific board, the General Advisory Commission. To a large extent, the GAC was the new scientific presence in Washington, but it was a part-time body, almost wholly absorbed in matters related to atomic weaponry.

Clearly there were policy and administrative matters that required or attracted the presence of scientists in Washington, both for the support of science and the employment of science for national purposes. But all along, those attracted or summoned there knew that their profession provided no credit for political accomplishment. Honor or advancement were accorded only for achievement at the laboratory bench, duly recorded in the appropriate professional journals, or in the direct administration of science. Time in Washington might further the career of a lawyer, journalist, or businessman, but in the scientific community an immutable criterion applied: What have you done lately in research?

Traditionally, the governing of science had been a part-time business. Except for a brief period in the 1920's, the presidency of the National Academy of Sciences was a part-time position until 1965—the 102nd anniversary of the Academy. Conant and Karl Compton retained their university presidencies throughout the war, despite deep involvement in the management of OSRD. Conant resided in Washington, but rarely missed the semimonthly meetings of the Harvard Corporation. Compton remained in residence in Cambridge. Aware of this peculiarity of his profession, Bush, in *Science, The Endless Frontier,* recommended that the government's traditional conflict-of-interest regulations be altered to reconcile the need for academic scientists to serve government with their preference to remain on campus.[23] "I would do everything possible to get the best men," he later testified, "and one thing that we can certainly do is to make it possible to utilize part-time men, for certainly there will be some men in the country whom we could get part-time and could not get full-time, and who will be just the type of men we ought to have." [24]

The government's lagging salary structure in part explained the reluctance to move from the campus to the federal bureauracy.* But the aversion to government employment, or even manning outposts of the scientific community in Washington, had roots far deeper than simply the desire to protect one's own financial interests. The laboratory and the campus provided a prestigious, comfortable enclave, one that was secured by rules of tenure and bounded by pleasant ceremonies and honors. It was within this enclave that the leaders had designed their careers, careers which began with arduous training, and which required rigorous apprenticeships and outstanding performance before recognition was achieved. Why should this be abandoned for the uncertainties and indelicacies of Washington? As one of the most eminent leaders of the scientific community, long involved in the uppermost echelons of science and government—but always with a private institution as his home base—explained, "When the hired hands appear before congressional committees, the congressmen can be a little bit rough. But I could always say, 'Well, for God's sake, if you don't want to hear me, well, the hell with it. I can always go back to my university.' "

---

* There is, of course, a great irony about government being unable to compete with the salaries that many universities and research organizations pay their own staff members, since it is the availability of federal funds that helps these private organizations to exceed the Civil Service pay scales. At present, the top pay for a government scientist is approximately $26,000 a year. A university-employed scientist, especially one who is qualified for the topmost government positions, not only often receives a basic salary in excess of that sum, but also may qualify for a government-subsidized salary for research performed during the summer months, plus consulting fees from government agencies and private industry. The former usually pay $75 to $100 a day; private firms often pay upwards of $250 a day.

But how did the scientific community look after its own affairs and also serve the needs of government in the first decade or so of the postwar collaboration? The answer is to be found in an incredibly intricate structure of interlocking relationships whose most notable characteristic was the ubiquity of its limited *dramatis personae.* "What official positions do you have with the government?" Rabi, chairman of the Columbia physics department, was asked in 1954 when he appeared before the security board hearing the case of J. R. Oppenheimer. "Let me see if I can add them up," he replied. "At present I serve as chairman of the General Advisory Committee [of the AEC] . . . I am a member of the Scientific Advisory Committee to ODM [Office of Defense Mobilization], which also is supposed to in some way advise the President of the United States. I am a member of the Scientific Advisory Committee to the Ballistics Research Laboratory at Aberdeen Proving Ground. I am a member of the board of trustees of Associated Universities, Inc., which is responsible for the running of Brookhaven Laboratory. I am a consultant to the Brookhaven National Laboratory. I was a member of the project East River [a study of civil defense], but that is over. I was at one time the chairman of the Scientific Advisory Committee to the policy board of the Joint Research and Development Board [in the Defense department], and a consultant there for a number of years. I am a consultant to Project Lincoln [a study of continental air defense]. . . . I added up what it amounted to last year, and it amounted to something like 120 working days. So you might ask what time do you spend at Columbia." [25] Rabi was in many ways extraordinary, but his ubiquity was a commonplace characteristic among his fellow statesmen of science. In 1957, for example, when Detlev Bronk, perhaps the most multi-hatted of science statesmen, appeared before a congressional committee, the chairman wryly greeted him with, "We could not operate without our distinguished friend, Dr. Bronk, chairman of the National Science Board, president of the National Academy of Sciences, and president of the Rockefeller Institute for Medical Research, and president of about ten other organizations." [26] According to a close and admiring colleague, Bronk had accepted the Rockefeller presidency with the intention of retaining the presidency of Johns Hopkins University (while he also held the presidency of the Academy)—a design that produced an anguished ultimatum from the Rockefeller board, with the result that Bronk departed from Hopkins to take the Rockefeller post. Bronk says the tale is apocryphal, but if it were in fact true, it would not have been extraordinary; rather, it would simply have been an extreme case of a commonplace practice. Concurrently, or in close succession, Lee DuBridge, while serving as president of Caltech, was a member of the General Advisory Committee of the AEC, the Naval Research Advisory Board, the Air Force Science Advisory Board, the President's Communications Policy Board, the National Science Board, the Science Advisory Committee of the Office of Defense Mobilization, and the National Manpower Council. Oppenheimer, who simultaneously chaired the General Advisory Committee of the AEC and served on the Science Advisory Committee, as well as some dozen other advisory groups, panels, and committees—while holding the directorship of the Institute for Advanced Study, at Princeton—provides an amusing insight into this fragmentation of roles. Following the first Soviet nuclear explosion, he told the security board, "I came down to Washington and met with a panel. I see it says here in my summary [notes?] that this was advisory to General Vandenberg. I was never entirely clear as to who the panel was supposed to advise." [27] *

Did not these interlocking appointments arouse concern as to conflict of interest? Not especially. When the advisory apparatus was in its early stages of development, little attention was paid to conflict of interest. It was taken as a matter of faith that the dispassionate ways of scientific inquiry could readily be transposed to the committee room where policy matters were weighed. In later years, under prodding from Congress and the press, the scientists began to pay more attention to conflict-of-interest hazards, but as late as 1967, John Coleman, Executive Officer of the National Academy of Sciences, wrote the following to Lawrence R. Hafstad, vice-president of research for General Motors, when Haf-

---

* For many, though certainly not all, of the leading figures in the advisory apparatus, the government-subsidized boom in advanced technology provided unusually lucrative opportunities. Thus, a laudatory profile of Jerome Wiesner notes that "In 1955, with Zacharias, Rabi, and several other physicists and engineers, Wiesner founded Hycon Electric, Inc., a firm that specialized in developing and marketing various types of electronic equipment and in taking on research jobs for the government. In due course, Wiesner became the chairman of the board. . . . In the late nineteen-fifties, too, working on a retainer basis, he served as a consultant to several other companies. His earnings from these ventures into the business world have enabled him to do various things he might otherwise never have contemplated." Among those cited were purchase of a summer home, financial assistance to students, friends, and colleagues, and acceptance of the relatively low-salaried post of science adviser to President Kennedy. From Daniel Lang, *An Inquiry into Enoughness, Of Bombs and Men and Staying Alive,* "A Scientist's Advice," McGraw-Hill, 1965, p. 109.

stad inquired as to the advisability of accepting an invitation to serve on the Academy's Committee on Undersea Warfare: "You raised the question of possible conflict of interest in view of your present employment with General Motors. We do not believe this is a serious problem for several reasons. First, it is well understood by the agencies we advise and the members of the Committee that they serve as individuals, rather than as representatives of their employers. . . . The Undersea Warfare Committee has always had some representation from industry; for example, the Bell Laboratories and the Electric Boat Co., both major contractors in the field. We have never detected any conflict nor, indeed, has the question been seriously raised by others."

All in all, the first postwar decade saw the creation of a most intricate maze for administering the collaboration of science and government. In 1950, when the long-gestating National Science Foundation at last came into existence, Alan T. Waterman, chief scientist of ONR, the "shadow" NSF, became director. By Act of Congress, the new agency for basic research was directed to "evaluate" and "correlate" the nation's scientific activities, as well as "to develop and encourage the pursuit of a national policy for the promotion of basic research and education in the sciences." [28] But from its day of birth NSF had neither the inclination nor the energy to impose its surveillance or will upon other scientific organizations. Waterman discloses that before taking the position, he exacted an assurance from the Bureau of the Budget that the fledgling foundation would not be required to fulfill a role that might bring it into conflict with far richer and politically more influential research agencies, such as the AEC and the Defense Department. In its fourth annual report, NSF frankly stated: ". . . it is clearly the view of the members of the National Science Board [NSF's top advisory body] that neither the NSF nor any other agency of Government should attempt to direct the course of scientific development and that such an attempt would fail. Cultivation, not control, is the feasible and appropriate process here." [29] Twelve years after NSF's founding, the policy and evaluation functions remained virtually unused, and were removed by a presidential reorganization plan.

In the year following NSF's founding, Bush's long-ignored proposal for a "permanent Science Advisory Board . . . to advise the executive and legislative branches of Government" [30] was, in part, accepted by Truman. A White House press release, dated April 20, 1951, announced the creation of a twelve-member Science Advisory Committee, within the Office of Defense Mobilization, which came under the Executive Office. But again, a pressing military motivation—in this case, the Korean War—provided the impetus for bringing into being the long-dormant design of the statesmen of science. The assignment of the committee was heavily weighted toward scientific matters of military significance, but it was not excluded from addressing itself to other aspects of science. Nevertheless, the committee chose, at least in its first years, to interpret its mandate conservatively. The first chairman, Oliver Buckley, came out of Bell Labs, which was at the institutional crossroads of basic research and sophisticated military technology. But Buckley, age sixty-three at the time of his appointment, was not part of the relatively youthful, war-tempered group of scientists and administrators that had begun to crystallize at the inner core of the science-government relationship. Furthermore, virtually every one of his colleagues on the SAC was deeply enmeshed in full-time professional responsibilities as well as a plethora of part-time committee assignments. Ostensibly, the newly formed SAC represented the peak of the pyramid in the science-government structure, but in its first few years, though its membership was influential, the committee itself—by the deliberate design of Chairman Buckley—was not. DuBridge, who succeeded to the chairmanship in 1952 when Buckley became ill, recalls that Buckley "announced to the National Academy of Sciences at one of their meetings that his conception of the purpose of this committee was that it was to maintain an overview of the status of science in America and of the scientists, and be alert to the needs of the government, especially in case of a new impending war or emergency. Then, when, as and if any new emergencies should develop, this group would have in the top drawer of its desk a plan of action and say, 'This, Mr. President, is the way we feel a new OSRD should be established in view of current circumstances, current people, and so on.' Buckley said at the time that this committee was not supposed to do anything, as he understood it, and that his objective was to keep the committee inactive except to meet occasionally to talk about this question of what we should do and in what ways was the OSRD no longer suitable to a new emergency if it should come now and who are the new people that have emerged that ought to be called on to come in. . . . And so it was inactive except in a vague and general way."

Speaking of the SAC, Oppenheimer later noted, "In the autumn of 1952 we had a two- or three-day meeting—probably two days—at

Princeton of this full committee to see whether we had any suggestions to pass on to the new [Eisenhower] administration as to the mobilization of science. I think we concluded that we had been of no great use and that as constituted and conceived we should be dissolved." Nevertheless, the committee felt that some device should exist to provide the National Security Council with "technical advice of the highest order" and "access to the whole community of science." Oppenheimer added that he and DuBridge discussed the matter with Nelson Rockefeller, who headed a committee to study reorganization of the Executive Branch. "We talked a good bit about our good-for-nothing committee . . ." Oppenheimer said. "We thought we were dead. We were, but not quite." [31] The committee remained in being.

Such was the "government of science" as the relationship between the scientific community and its newfound patron, government, coalesced following World War II. Let us now look at that "government" in some detail and then go on to examine a series of episodes that illuminate the intricacies of the politics of pure science.

NOTES

1. Meg Greenfield, "Science Goes to Washington," *The Reporter,* September 26, 1963, pp. 20–26.
2. James B. Conant, *Modern Science and Modern Man* (Garden City, N.Y.: Masterworks Program, 1952, by arrangement with Columbia University Press), p. 116.
3. *In the Matter of J. Robert Oppenheimer,* Transcript of Hearing before Personnel Security Board, Atomic Energy Commission, 1954; see Latimer's testimony, pp. 658 ff.
4. "Plea for Closer Look at U. S. Security," *Christian Science Monitor,* May 5, 1950; statement signed by Duncan S. Ballantine, McGeorge Bundy, Martin Deutsch, J. K. Galbraith, William R. Hawthorne, John E. Sawyer, Arthur M. Schlesinger, Jr., Charles H. Taylor, Jerome B. Wiesner, Jerrold Zacharias.
5. J. R. Marvin and E. J. Weyl, "The Summer Study," *Naval Research Reviews,* August 1966.
6. Nick A. Komons, *Science and the Air Force, A History of the Air Force Office of Scientific Research,* Historical Division, Office of Aerospace Research (Arlington, Va., 1966), p. 56.
7. *Senate Independent Offices Appropriations, Hearings,* 1956, p. 407; from compilation supplied by the National Science Foundation.
8. Charles V. Kidd, *American Universities and Federal Research* (Cambridge, Mass.: Harvard University Press, 1959), p. 235.
9. War Department, Office of the Chief of Staff, memorandum, "Scientific and Technological Resources as Military Assets," April 30, 1946.

10. Leslie R. Groves, *Now It Can Be Told* (New York: Harper and Brothers, 1962), p. 374.
11. R. G. Hewlett and O. E. Anderson, *The New World, 1939/46*, Vol. 1, *A History of the United States Atomic Energy Commission* (University Park, Pa.: Pennsylvania State University Press, 1962), p. 628.
12. *Ibid.*, p. 635.
13. George Rosen, "Patterns of Health Research in the United States," *Bulletin of the History of Medicine*, Vol. XXXIX (May-June 1965), p. 220.
14. *Department of the Navy Appropriations Bill for 1949*, House of Representatives, p. 959.
15. "The Evolution of the Office of Naval Research," by the "Bird Dogs," *Physics Today*, August 1961, p. 30.
16. "The Scientists: A Group Portrait," *Fortune*, October 1948, p. 166.
17. *Bulletin of the Atomic Scientists*, August 1947, pp. 221–30.
18. *Ibid.*, p. 222.
19. *Security, Loyalty, and Science* (Ithaca: Cornell University Press, 1950), p. 22. Gellhorn's figures are derived from the AEC's *Fifth Semiannual Report* (1949).
20. Quoted in *Physics Today*, August 1961, p. 33.
21. D. S. Greenberg, "The National Academy of Sciences: Profile of an Institution," three-part series, *Science*, April 14, 21, 28, 1967.
22. *Hearings on Science Legislation*, Subcommittee of the Senate Committee on Military Affairs, 1945, p. 430.
23. Vannevar Bush, *Science, The Endless Frontier* (1945; reprinted by National Science Foundation, 1960), p. 38.
24. *Hearings on Science Legislation*, 1945, Part 2, p. 214.
25. *In the Matter of J. Robert Oppenheimer*, p. 451.
26. *House Independent Offices Appropriations for 1958, Hearings*, p. 1270.
27. *In the Matter of J. Robert Oppenheimer*, p. 75.
28. National Science Foundation Act, Public Law 507, Section 3, 81st Congress.
29. National Science Foundation, *Fourth Annual Report*, p. viii.
30. Bush, p. 7.
31. *In the Matter of J. Robert Oppenheimer*, pp. 93–94.

BOOK THREE

VIII
The Government of Science

One of the . . . features of the federal system [for supporting scientific research in universities] . . . is that it is now characterized by a complicated advisory network—in fact, too complicated to describe and which magazine articles have made fun of.

> DONALD F. HORNIG, Presidential Special
> Assistant for Science and Technology,
> in *Science and the University*, 1966

There is a current saying among government supporters of research that scientific research is the only pork barrel for which the pigs determine who gets the pork.

> KENNETH M. WATSON, Physicist, "A Comment on the Motivation
> for Studying Elementary Particle Physics,"
> in *Nature of Matter*, 1965

The most influential statesmen of pure science systematically assert that there neither is nor should be a national policy for science; and, further, that science neither has nor should have any central directive mechanisms, let alone anything that might be considered a "government." In fact, Jerome Wiesner once wisecracked that one of his functions as Kennedy's science adviser was to "protect the anarchy of science." And, when the panorama of basic science is observed from certain perspectives, it must be acknowledged that Wiesner and his colleagues are correct, both in description and prescription. In spirit and in practice, laissez faire pervades the scientific enterprise, with apparent benefit for the conduct and progress of science. Laissez faire is not the universal rule, but it is so widely prevalent that, more often than not, the initiative for scientific inquiry rests with the individual researcher. He decides what he wishes to investigate, and once the merit of the subject and his qualifications to deal with it have been certified by his peers, he is essentially on his own. To that extent, pure science is governed by its

151

own internal standards of scientific value, rather than any authoritative superstructure.*

However, what must be joined to this recognition of the freedom that does prevail in pure science is the realization that the laissez-faire system takes place within the bounds of an intricately constructed, subtly functioning system of government that, in effect, defines the possibilities of science by governing the availability and use of resources. If the system is not bound by rigid policy, it is nonetheless bound by well-rooted practice and custom. If its principal political objective is the maintenance of independence, it is an objective that is pursued within limitations imposed by the various authorities that provide funds. And, finally, though it may appear that affluence and anarchy are the prime characteristics of the contemporary scientific community, the fact is that economic constraints have never been absent from the affairs of science, and that, in response to these constraints, processes of decision-making, implemented through old or newly established institutions, have come into being. In fact, there *is* a government of science, but, like so much else associated with the scientific community, it is not easily revealed or understood. Nevertheless, let us now proceed to look at it in some detail, taking the approach that one route to understanding the institutions and *modus operandi* of any government is to examine the manner in which it perceives, confronts, and attempts to deal with any problem that it finds difficult and important. For our purposes, a recent problem involving the field of chemistry will do nicely, for in

---

* Scientific laissez faire generally finds its most hospitable setting in an academic environment, where the payoff from research is measured in terms of scientific papers. In industrial laboratories, the spirit of scientific independence must reckon with the profit motive. It is often difficult to mesh the two, but industry has become quite sophisticated at meeting the researcher halfway. For example, in an interview, Arthur M. Bueche, head of General Electric's Research and Development Center, was asked, "After you have chosen an area that looks fruitful, how do you steer the laboratory toward it?" Among the methods he described was the following: ". . . We try to hire people from outside who are interested in the area we have chosen and hope they will come up with new ideas." ("Making Research Pay," *International Science and Technology*, February 1967, p. 74.) Or, consider this recruiting advertisement by an electronics firm: "AVCO Everett: You'll think you're doing research in a university. . . . That's the difference between AVCO Everett Research Laboratory and other industrial laboratories. At AVCO Everett the atmosphere is one of complete freedom . . . freedom so necessary for the highest degree of creative thinking." (*Science*, March 10, 1967, p. 1344.)

following that problem, we will be conducted on an intimate exploration of the government of science.

Each branch of science possesses substantive characteristics that generate particular economic and political problems for its practitioners. For example, high energy physics, which will later be closely examined, is one of the most esoteric of scientific fields, and, without any doubt, also the most expensive, a combination of characteristics which has proved to be particularly burdensome in this science's relations with other fields and with the federal agencies that provide financial support for research. Chemistry has quite different peculiarities. Far from being esoteric, it is the most pervasive and probably the most widely understood and useful of all the sciences, as evidenced by such disciplinary amalgamations as biochemistry, geochemistry, physical chemistry, and chemical physics. Growing out of these hybrid disciplines are a score of subspecialities, such as petroleum chemistry, food chemistry, and clinical chemistry. (It might be added that the chemistry set is perhaps the most popular of scientific toys.) Of all the sciences, chemistry has the closest linkage to industry, though since head-counting in the sciences is a highly uncertain business, it is difficult to assess the deployment of the profession with any great precision. According to a study conducted under the auspices of the National Academy of Sciences, professional practitioners of chemistry in this country—at all levels of training—numbered 125,000 in 1965; of these, 100,000 were employed by industrial firms. Of the approximately 23,000 chemists engaged in basic research, some 12,000 were in universities, 8000 in industrial laboratories, 2000 in government laboratories, and the remainder in private research institutions.[1] The National Register of Scientific and Technical Personnel lists 63,053 chemists, with nearly 38,000 of these employed by business or industry. In any case, as far as place of employment is concerned, the contrast with other disciplines is striking. The Register lists 27,135 in the biological sciences; of these, 15,872 are in educational institutions. It lists 26,698 in physics, with 11,611 of these in educational institutions. Thus, in sheer numbers, chemistry is the most heavily populated of the disciplines, and its practitioners are to be found in relatively large numbers along the entire spectrum of research and development, with their institutional settings ranging from major industrial laboratories to small departments in minor colleges.

For the purpose of obtaining financial support from the federal government, which is overwhelmingly the largest single source of funds for basic research in chemistry, there are obvious advantages to this far-flung dispersion. Unlike those disciplines that are heavily dependent upon a single federal agency—such as medical research is dependent upon the National Institutes of Health, or physics upon the Atomic Energy Commission—chemistry is woven into the programs of every federal agency concerned with science and technology and it taps into the budgets of each of them.

Paradoxically, this dispersion of support was the cause of a serious money problem that university-based chemistry began to feel in the early 1960's. To the extent that chemistry research helped achieve utilitarian objectives—such as insect control or the development of nuclear fuels—it was relatively well supported, not for the sake of chemistry, but for the purpose of getting the job done. By definition, then, this support was for applied or developmental research in chemistry, relatively little of which was conducted in universities. As for basic research, most of the agencies that employed chemistry in working toward identified objectives also devoted a small portion of their overall research budgets to basic research in chemistry, either conducted by their own employees, through awards to universities, or both. But, in the grand scheme of federal support for research, the main financial prop for chemistry for the sake of chemistry was supposed to be the National Science Foundation, whose *raison d'être* is the support of basic research for no other purpose than the advancement of scientific understanding and education. NSF's responsibility, however, is not confined to chemistry. By the Act of Congress that brought it into existence, NSF was made responsible for the "mathematical, physical, medical, biological, engineering, and other sciences." And, as it turned out, with these far-ranging responsibilities, NSF was not spending very much on chemistry. In fiscal 1962, Congress had appropriated $263 million for the Foundation, more or less leaving it to NSF to decide how it would divide this sum among its various clients.* The Foundation, in

* Though Congress as a whole and through its committees has in recent years become increasingly attentive to the affairs of federal agencies that support basic research, the congressional interest has rarely if ever got down to the level of expressing preference for one field of basic research versus another. Rather, the Congress has concerned itself with pork-barrel issues such as the geographic distribution of research funds, problems of emphasis on research adversely affecting teaching, and the linkage between basic research and practical applications. NSF, which is annually raked over by the appropriations subcommittees

turn, allocated $8 million to universities for research in chemistry and the purchase of chemistry research instruments. The rest of NSF's budget went to other fields and programs. The following year, NSF's budget rose to $322 million; the amount for chemistry went up to $9.5 million. In fiscal 1964, the NSF budget went up to $353 million; of this, chemistry received a sliver totaling $10.5 million.[2]

Now, in looking across the whole array of federal research agencies, it is important to note that since no single granting agency considered chemistry to be its central concern, there was no agency that deliberately sought to be responsive to the needs of chemistry, as, for example, the AEC, with an assortment of carefully assembled advisory committees, attunes itself to the needs of physics. As for NSF, it included chemists in its various advisory mechanisms, but it included scientists drawn from every other discipline, too, and, as a result, basic research chemists had no special influence in NSF, the agency upon which they felt most dependent for general support.*

It was against this background of having many places but no particular home in government that some leading practitioners of chemistry began to feel that the flow of federal money was not keeping pace with the aspirations they held for their profession. In 1961, Professor William Doering, a Yale chemist, undertook a survey of graduate chemistry departments to obtain information about their needs for items of equipment costing over $2000 (which, as we shall see, is a modest price in the marketplace for modern research equipment). As is invariably the case when such "shopping list" surveys are compiled, the results showed unmet needs of prodigious proportions. A report based on Doering's survey was presented to the twenty-four-member National Science Board, which is NSF's top advisory board, but there is no evidence that it had any effective impact. In October 1962, John D. Roberts, chair-

---

that handle its budget, is yet to be criticized, let us say, for its relative allocations between astronomy and biology. Congressmen are willing to leave such matters entirely to the scientists who administer the agencies.

* This is not to suggest that chemists have been underrepresented in the upper advisory councils. Two presidential science advisers, Kistiakowsky and Hornig, come from the ranks of chemistry, as did Glenn T. Seaborg, chairman of the AEC. Nevertheless, it was neither their duty nor choice to function as representatives of their disciplines and chemistry did not derive any particular benefit from the presence of three of its members in these elevated positions. In fact, if anything, it is curious that with many chemists occupying influential positions in the advisory apparatus that links science and government, chemistry quietly slid into what many of its leaders sincerely came to regard as a serious financial crisis.

man of the division of chemistry and chemical engineering at Caltech and chairman of NSF's advisory committee on the mathematical, physical, and engineering sciences, wrote to the board that the percentage of the NSF budget going to chemistry had dropped from 47.2 percent in 1952 to 8.6 percent in 1962. He added that ". . . the drop in this figure has occurred at just the time when there has been a great increase in receipts of meritorious chemistry [research] proposals . . ." * Again, there is no evidence that NSF chose to pay any particular attention. It was not that the Foundation was indifferent to chemistry; rather, the plaints of the chemists were commonplace. No field of research ever felt that it had enough, and for those who had to make the decisions on what to support, there was no easy method of distinguishing the cries of the painfully deprived from those of the affluent who merely sought more icing on their cake.†

When 1963 began, the chemists still lacked relief. From personal experience and from conversations with colleagues, they knew that the available funds were inadequate to support a significant portion of the research that they considered worthwhile. But their grievance had not yet crystallized. What it amounted to was a general feeling of discontent, backed up by a few statistics. But the feeling was strong, and it grew stronger as two factors—one technological, the other political—began to merge in 1963 with painful consequences for the scientific community in general, but for the politically rootless chemists in particular.

To comprehend the technological factor it is necessary to note that science is, indeed, an "endless frontier," for each question that is successfully answered almost always brings into view new and unexplored

---

* The striking difference between the two percentages is explained by the fact that in 1952, NSF was barely two years old, and had not yet established any of the major programs that were later to absorb large portions of its budget. Compared with oceanography or radio astronomy, for example, chemistry is a relatively inexpensive brand of research. Though year by year, NSF increased the number of dollars for chemistry, the amounts, as a proportion of the overall NSF budget, actually declined when the Foundation moved into various fields of "big science."

† The problem is not a peculiarity of *American* science; rather, it is a peculiarity of science, regardless of locale. Nikita Khrushchev is quoted as once having said, in response to a complaint of inadequate support for research, "You must never believe all these things which the scientists say because they always want more than they can get—they are never satisfied." (Quoted in *Nuclear Scientist Defects to United States*, Senate Internal Security Subcommittee, 1965.)

questions.* And at the heart of this amoeba-like proliferation of scientific questions is the amoeba-like proliferation of instrumentation. Since the advent of electronics, the designers and manufacturers of scientific research equipment had been performing splendidly, bringing to a close the era in which, as Philip H. Abelson described it, "Operations in the chemical laboratory were not much different from those conducted in the kitchen." [3] Drawing upon technology whose development had been initiated or accelerated during World War II, the manufacturers, year after year, brought forth research instruments of such great power and versatility that any laboratory not equipped with a good array of these devices was severely handicapped in the rigorous competition that characterizes scientific research. Equipment is no substitute for creativity and skill, but without the equipment, even extraordinarily creative and skillful researchers are like horse-and-buggy drivers in a motorized Grand Prix competition. And, year by year, the equipment became ever more complex—and ever more costly. The popular image of the chemist depicts him as working with a few test tubes, flasks, and a Bunsen burner. A small proportion of chemists still work that way. But on the frontiers of chemical research, where, naturally, most would prefer to be, the chemist is surrounded by a collection of indispensable instruments that cost a small fortune. Today, as Abelson notes, "Gas-liquid chromatography permits quick and effortless identification and measurement of twenty or more substances present in a tiny droplet of a sample. In an earlier day, to get comparable information might have required gallons of material and weeks of work." [4] Such equipment is relatively inexpensive—around $15,000 for a complete outfit. More costly are the various types of spectrometers that permit the rapid measurement and analysis of infinitesimal quantities of materials, with a precision that literally is as fine as individual atoms and molecules. A stripped-down version of the device known as the nuclear magnetic resonance spectrometer costs $30,000; more elaborate models cost upwards of $50,000. A mass spectrometer, which is a commonplace piece of

* Until a few years ago, for example, it was an article of faith in science that the element xenon was so thoroughly inert that it could not be combined with any other substance. Then it was discovered that xenon could be made to react with fluorine. Almost immediately, scientists throughout the world rushed into research on xenon. Typically, the discovery of xenon's capacity for interaction answered one question and simultaneously created many others. After about two years of research, it appeared that xenon's basic properties had been thoroughly explored, and the chemists turned elsewhere.

equipment in any self-respecting modern chemistry laboratory, typically costs approximately $60,000; there are models that run as high as $100,000. According to one estimate, these devices, as well as others that have become the mainstays of modern chemical research, lead to a cost of approximately $750,000 to equip a good moderate-size graduate chemistry department.[5] To this capital investment must be added $100 to $500 an hour for computer costs, plus, of course, salaries, supplies, and an assortment of odds and ends that are necessary for the functioning of modern science. Among these are travel funds to attend conferences to report the scientific findings and discuss the new questions that are produced by the laboratory's arsenal of modern machinery.

Thus, the chemists faced the spiraling problem of success making their craft ever costlier. They were not alone in this situation, of course, for the same problem confronted physicists, space researchers, medical researchers, and other scientists. Where they were alone (except for the earth scientists, among whom a similar problem helped create the Mohole debacle, discussed in Chapter IX) was in the administrative peculiarity that left them without any focal point of representation among the federal agencies that support research. And it was against this background that the growing costliness of their profession collided with a politically induced factor, namely, a sudden tightening of federal support for research.

The fiscal year 1949 produced the first peacetime budget in which federal expenditures for research and development passed the billion-dollar mark. As noted earlier, some 90 percent of this money was for the purpose of developing hardware, mostly military hardware. But a bit of that development money, and most of the remaining 10 percent, went directly to universities, principally to their basic research departments. And then year after year, the volume of federal R and D funds grew, often in great leaps. By 1957, the amount had risen to $4.4 billion; in 1962, it passed the $10-billion mark. Two years later, the figure was $14.8 billion. This gusher of money produced innumerable effects, but among them, the two that principally concern us were these: science and technology became attuned to a rhythm of ever-growing financial support, and Congress became increasingly interested in, and querulous about, this suddenly emerged giant in federal affairs. Typical of congressional sentiment was the declaration, in 1963, of Representative Joe L. Evins (D-Tenn.): ". . . The scientist," he said, "does not know the value of a dollar." [6] Evins made the assertion in the course of hearings at which NSF, whose appropriations had started with $225,-000 in 1951, was seeking for fiscal 1964 the sum of $589 million—an increase of $267 million over its previous annual budget. Congress, however, would cooperate only to the extent of increasing the budget by $30 million—a superb gain by prewar standards, but a mere crumb in the era of postwar affluence. All along the line, Congress, after seeing funds for research and development triple in a period of seven or eight years, was in a sour mood about this incomprehensible, burgeoning monster that, it now felt, it had absentmindedly nurtured. NSF officials argued that the bulk of the growth was for space and military hardware. Kennedy had inherited a space budget of $741 million; by 1964 it was over $4 billion. Most of the federal R and D money goes for developmental work, the scientists argued. But few congressmen noted the details. Slowly the heretofore ever-growing gusher was being restrained. Because of the sudden surge of space money, overall federal funds for research and development rose from $11.9 billion in fiscal 1963 to $14.8 billion in 1964. But once the space program was established in the budget, at an approximate annual level of $5 billion, the rate of growth suddenly shifted. As pictured on a graph, the decade-long sharply upward line now leveled off to a position just slightly above the horizontal. Looking at the postwar history of federal support for research and development, William D. Carey, assistant director of the Bureau of the Budget, prophetically declared in a speech that "the justification for the sixteenth and seventeenth billion will have to be very different from the justification for the first billion." [7] From then on the money was to come hard.

Now, what effect did this politically induced financial turn of events have upon the chemists? It filled them with painful anxieties for the future, as can be seen in an excerpt from a letter that a Caltech chemist, Carl Niemann, chairman of the Academy's chemistry section, wrote to a number of colleagues in mid-1963: "There is little doubt that we are faced with the prospect of a leveling off of government support in all of the basic sciences. Unfortunately, increases in support for basic research in chemistry have been smaller than in other disciplines during the last few years. This means that the support available for established branches of chemistry is likely to level off at an undesirably low level which would mean a dangerous curtailment of expansion into new areas"—which brings us, in our tour of the government of science, to that curious creation of the National Academy of Sciences—the Committee on Science and Public Policy, conveniently referred to as COSPUP.

COSPUP was the creation of one of the grand and prescient statesmen of science, George B. Kistiakowsky, who, while serving as presidential science adviser in the last two years of the Eisenhower administration, foresaw that the time would soon come when government would ask very hard questions about levels of support for basic scientific research. It cannot be said that science was granted a blank check prior to 1960. Far from it. But, though government did not grant the scientists all they sought, it nevertheless granted a great deal. The important point, however, is that what was given, was given on faith in the value of science, and not as a consequence of any systematic assessment of the place or value of science in national affairs. What Kistiakowsky foresaw was that sooner or later, faith would be supplanted by an insistence upon the scientific community offering something more substantial than its traditional argument that it is in the national interest to support science and leave it alone. Thus was formed COSPUP, a twelve-member committee, chaired by Kistiakowsky, and holding a mandate to study and pronounce upon any problem that might come under the boundless heading of science and public policy. COSPUP held the unique status of being the only one of the thousand or more Academy committees and panels on which only Academy members might serve. On all the others, non-Academy members were brought in to serve under the auspices of the Academy's operating arm, the National Research Council. COSPUP, too, could employ the services of non-Academy members, but the finished product was always issued in the name of COSPUP. This may sound like a distinction without a difference, and in many respects it is just that. But to the extent that there could be a single voice for that amorphous body known as the scientific community, it was COSPUP's.

What Kistiakowsky was particularly interested in was that COSPUP should serve as a scholarly, dignified advocate for research by producing inventories of the scientific status of various fields and assessing the resources required to follow promising lines of inquiries. It was his belief that persuasive argument was the best weapon available to the profession. Science, he argued, could not make its way in the public process by employing conventional lobbying techniques; it had neither the numbers, the votes, nor the money that were regularly employed by other governmental supplicants. Furthermore, Kistiakowsky felt, the rough and tumble of conventional politics was incompatible with the ways of science. "I thought it would be terribly bad if we tried to get support through outright lobbying. I would not be a part of it. I felt we should do it through eloquence—eloquence based on the facts." And that is how COSPUP sought to function—with eloquence based on the facts. Some may feel that this is a naive approach to the problem of commanding the resources of the government of the United States. But, first of all, it is difficult to perceive any attainable alternative for that relatively tiny segment of the electorate that is made up of basic scientists. Then, too, it is worth noting, that tiny segment has not done too badly. Finally, it is also worth noting that Kistiakowsky had arrived at an informal understanding with Kennedy's science adviser, Jerome Wiesner. When COSPUP came forth with a study, it was agreed, the White House science advisory apparatus would seriously consider the data and recommendations. Since Kistiakowsky was a member of the President's Science Advisory Committee (PSAC), and the staff serving PSAC had close ties to the Bureau of the Budget, the informal Kistiakowsky-Wiesner agreement, far from being naive, had a good deal of practical significance in the political affairs of science. In a sense, the arrangement was not too different from having a loan applicant at a bank serve as adviser to the loan officer handling his application.

When the Academy held its annual meeting in April 1963, COSPUP was a tender one-year-old, and Kistiakowsky was gently urging his colleagues to employ this fledgling for identifying and justifying their claims for federal support of their fields of research. At the same time, John D. Roberts, the Caltech chemist, was seeking to rouse his colleagues to the financial plight of chemistry. In this, as in so many other instances, the traditional practice of interlocking appointments served a useful purpose. Roberts viewed with concern not only from his position as a member of the Academy's chemistry section, but also as one of NSF's top advisers. In this latter capacity he was in an excellent position to see how chemistry was faring in NSF's budgetary decisions. Kistiakowsky viewed the scene as a chemist, chairman of COSPUP, and member of the President's Science Advisory Committee. Since *ad hoc* committees are a way of life in the scientific community, it was decided at that April meeting to establish a temporary group to conduct a reconnaissance to determine what would have to be done to examine in detail the financial plight of the profession. The chairmanship of the *ad hoc* group was filled by Frank Long, of the Cornell chemistry department, who was also a member of PSAC. Not long afterwards, NSF director Leland J. Haworth asked for a copy of Roberts' remarks to the April meeting of the Academy's chemistry section; Haworth explained that AEC chairman Glenn Seaborg, a member of the chemistry section, had brought Roberts' talk to his attention. NSF, it should be pointed out, was the source of financial support for COSPUP's operations. Meanwhile, Melvin Calvin, a Nobel laureate chemist, who was also a member of PSAC and of COSPUP, was urging his colleagues in the chemistry section to give serious thought to working closely with COSPUP. It was stressed that Kistiakowsky had emphasized that Wiesner had informally agreed to give serious attention to COSPUP's reports.

Thus, the seemingly inchoate government of science was stirring itself in response to the chemists' sense of maltreatment. On October 22, 1963, Long's *ad hoc* committee tentatively concluded that only a massive, in-depth study could accurately assess the financial situation. Meanwhile, the *ad hoc* group was comfortably coming into the embrace of COSPUP. Following the meeting, Long wrote to his colleagues, "In view of the committee's conclusions, it seems appropriate for us to accept Kistiakowsky's kind offer and forward any charges for travel for the October 22 meeting to his Committee. . . ." At the same time, Long reported that he and another PSAC member, Donald F. Hornig, chairman of the Princeton chemistry department, who was to succeed Wiesner as presidential science adviser, had made arrangements to take up the chemistry situation with NSF officials. The *ad hoc* committee reassembled in February 1964 and decided to make a study in the COSPUP manner. Absent from that particular meeting was Frank Westheimer, a colleague of Kistiakowsky on the Harvard chemistry faculty. As is occasionally the practice among the veteran committeemen of science, Westheimer was accordingly elected to head a committee, under the auspices of COSPUP, to make a comprehensive study of the state of American research in chemistry. He promptly received acceptances from fourteen chemists, all members of the Academy, to join him in making the study. NSF and the American Chemical Society provided financial support. And thus was born what is to date the most definitive study of the scientific and economic condition of any field of scientific research.

Covering 222 pages of a volume titled *Chemistry: Opportunities and Needs,* the Westheimer survey is a model of the eloquence desired by Kistiakowsky. Inevitably, it concluded that in chemistry, great scientific opportunities were being thwarted by unmet financial needs. The product of the Westheimer survey was a striking example of advocacy attired in the dispassionate, nonpolemical methodology of science. But the case it made was a compelling one, so compelling, in fact, that the government of science could not help but listen—though whether it acted in accordance with the Westheimer prescription is a separate matter, to be examined later.

Though basic researchers often gag on the proposition that utility should be the justification for public support of their work, there is no doubt that utility is the most palatable rationale for public subsidy—especially when basic research can be easily related to something as unarguable as the $40-billion-dollar-a-year sales record of the chemical industry. Mindful of this close linkage between basic research and a big money-making industry, the Westheimer survey set out to reconcile utility and scientific freedom by demonstrating that free, undirected basic research in chemistry was the source of much of the industry's prosperity. The technique for establishing this was relatively simple. First the committee compiled a list of forty post-1946 chemical developments of great commercial and industrial value—among them various drugs, plastics, fibres, and other synthetic materials. Then it collected the scientific and technical papers which first announced these developments. Since the traditions of publication require that reference be made to all contributory past research, the committee could easily assemble what, in effect, was a scientific history of the forty items on its list. What emerged from this history was that papers published in basic research journals predominated as the principal source of discoveries that led up to the development of the products under study. In the case of industrial products, 67 percent of the citations referred to articles in fundamental journals; for pharmaceutical products, the figure was 87 percent.* Thus, it could be argued that fundamental chemistry was an

* The Westheimer findings raise some interesting questions about Project Hindsight, the Defense Department study referred to in Chapter II. Hindsight, which was conducted mainly by engineers, concluded that very little postwar basic research had found its way into contemporary weaponry. There are several possible explanations for this contrast with the Westheimer findings: The COSPUP chemists, being basic researchers, were motivated to look for the industrial fruits of basic research; on the other hand, the Hindsight engineers, not being particularly

indispensable part of economically profitable chemistry. An economist might contend that a firm case could not be made unless it were demonstrated in each instance that the investment in basic research produced a return that was superior to any other feasible alternative, but that is slicing the matter a bit close for a group of scientists who simply wished to establish that basic research produces a big payoff.

Having established a compelling case for the great economic value of basic research in chemistry, the Westheimer group next applied itself to the task of determining the sum annually expended in this country for support of such research. In this instance, semantic problems and varying bookkeeping practices posed a substantial difficulty. Since chemistry is the most pervasive of disciplines, and disciplinary boundaries have become increasingly meaningless, it can be argued that traditional budgetary designations do not accurately reflect what is actually going on in the laboratory. The Westheimer group chose a simple and advantageous solution: chemistry, it stated, will be defined as what goes on in university chemistry departments and other research facilities that are predominantly staffed by holders of advanced degrees in chemistry. By employing that definition, they instantly excluded a vast amount of research that might reasonably be considered to be in the field of chemistry, and emerged with the conclusion that, relatively speaking, not much was being spent on this economically valuable science. The actual compilations add up to a morass of qualified and annotated numbers, but the principal finding was that "The grand total of expenditures (explicit and hidden *) for basic research to chemistry in U.S. universities in 1964, excluding construction, was thus about $90 million, with less than $60 million from federal sources." It was estimated that the federal government spent another $57 million for basic chemistry research in

---

concerned or knowledgeable about basic research, quite possibly were not on the lookout for its manifestations in the weapons systems they studied. Another possibility is that the managers of defense systems have simply been laggard in applying basic research findings, while the chemical industry has been extremely active.

* Support that is "hidden" includes such items as the overhead allowance that federal agencies provide to help pay the on-campus costs that are generated by the presence of federally supported research programs. Among these are clerical and janitorial services, increased demand on library facilities, and central administrative supervision. Overhead is an incessant source of squabbling between the federal agencies and the recipients of their grants, but there is no doubt that (1) virtually all universities are strapped for money, and (2) when a major research program is set up on a campus, its costs go beyond the mere performance of the research specified in the research application.

its own laboratories, and that the chemical industry reported basic research expenditures totaling $150 million, of which the federal government provided somewhere between $25 and $40 million.[8]

Now, since the motivation for the study was the belief that the available money was inadequate, and thirteen of the fifteen members of the committee were employed by universities, the committee largely confined itself from that point on to demonstrating that the support of chemistry in universities was dangerously below what was needed.

Since the question of what is enough is a difficult one, the committee approached the task from a number of directions. First, having established that chemistry is of great economic value, it furnished figures showing that in cost per Ph.D. produced, chemistry was one of the least expensive of all the sciences. In 1962, the report stated, the federal government obligated (that is, made available for eventual expenditure) $70.3 million for chemistry; in that same year, 1182 Ph.D.'s were granted in chemistry. By contrast, the figures for physics were $228.7 million and 735 Ph.D.'s. In the earth sciences, which includes oceanography, with its costly seagoing laboratories, $254.7 million was spent, and 252 Ph.D.'s were produced. This comparison of disciplines, costs, and production of doctorates was essentially a self-serving exercise in the meaningless matching of numbers. But since it came out favorably for chemistry, the chemists saw a purpose in including it. (When physicists not long afterwards prepared a similar study of their discipline, they dealt very gingerly with the fact that in cost-per-man, physics is very expensive.)

Next the study moved to establish that this valuable, inexpensive field of research was receiving relatively shabby treatment when it sought funds in Washington. The means of demonstrating this was simply to show how chemists fared in comparison with other scientists in obtaining money from federal granting agencies. The figures assembled by the chemists showed that their profession was the poor relation of American science. In fiscal 1963, it was stated, chemists sent 721 proposals to NSF, asking for a total of $56 million; NSF accepted only 240 of these proposals, and so sharply reduced the amount in each of them, that the total of funds actually granted came to only $9.5 million. In comparison, the study went on, all the other sciences put together sent to NSF a total of 4144 proposals, for a total request of $341.4 million; 2237 of these were accepted, and NSF paid out for them a total of

$90.1 million.* The result was that chemistry received only 17 percent of the dollars it sought, while other sciences received 26.4 percent. In the other federal agencies that support research, chemistry fared no better, the chemists reported. The Air Force Office of Scientific Research gave the chemistry applicants only 13.6 percent of the money they sought, while giving all the other sciences combined a total of 21.4 percent. As for the AEC, it might be concluded from the chemists' figures that their profession had been declared unwelcome by that multibillion-dollar research agency. The AEC, they reported, gave them only 5.1 percent of their requests, while giving 49.3 percent to the other sciences.

Thus, the chemists had established, at least to their own satisfaction, that their profession was economically valuable, relatively inexpensive, and poorly treated. Now what they had to establish was that they really could make good use of substantially greater support. This proved to be no more difficult for them than it would be, let us say, for a housewife to elaborate how she might effectively make use of a good deal of money for redecorating the living room.

First the study carefully detailed a vast inventory of problems—ranging from cancer to defense against chemical and biological weapons—for which it might reasonably be assumed that a better understanding of fundamental chemical processes would provide at least part of the key to solutions. Then it went on to note that while chemistry in a sense was booming (in a few years, the number of departments granting the Ph.D. in chemistry had risen from 125 to 140), the flow of money was not keeping pace with the growth. Questionnaires sent to university chemistry department chairmen throughout the country brought back the almost unanimous reply that funds were not available to provide equipment for rising enrollments; young Ph.D.'s were experiencing unusual difficulty in obtaining research funds, and, in general, a scarcity of money was restricting operations and limiting opportunities. After putting together these findings, projecting enrollments, assessing the equipment costs, computer purchase and rentals, and various other items, the Westheimer survey emerged with the conclusion that opportunities in chemistry had so outstripped financial support that compensatory treatment was in order. "Funds for basic research in chem-

---

* NSF's total expenditures were, of course, far in excess of $90.1 million in that year. The figures here refer only to grants awarded for individual research projects.

istry have grown by about 15 percent a year in the past decade," the report stated. "Under the stimulus of this financing, the science has become more vigorous than ever before. Our survey has led us to conclude that research opportunities and manpower needs in chemistry warrant a more rapid growth in funding." Though Kistiakowsky and others had come out just a few months earlier with the figure of 15 percent annual growth as a rule of thumb for science budgeting,[9] the Westheimer report—proceeding from the premise that chemistry had long been undersupported—recommended that support for chemistry in universities be increased annually by 20 percent for three or four years before reverting to the 15 percent level.* Since the federal government is the largest and most affluent of supporters of basic research, it was suggested that its contribution to chemistry in this period actually should be increased by 25 percent a year. Thus, from a level of $60 million in 1965, the federal sum would rise to $120.9 million by 1968. "We are convinced," the Westheimer committee stated at the conclusion of its report, "that our recommendations, if put into effect, will increase the vigor of chemistry, and will greatly benefit the nation."

Now, what happened after the report was delivered to the federal budgeteers whom it was intended to influence? Two years later Westheimer's own assessment was simply this: "The effects are difficult to assess." And indeed they are, for overall, the federal budget is so complex, and research budgets comprise so rococo an appendage to this complexity, that at any given moment it is difficult, and at times impossible, to determine how much is being spent on what. There is no doubt that the publication of the report was followed by an increase of federal support for chemistry. But why is another matter, since funds for chemistry had annually been on the increase before the report was even conceived. Nevertheless, a few important elements emerge with clarity. The first is that, though ultimate power to control events does not rest in the advisory structure of the science and government rela-

* For a few years, 15 percent was the magic number, generally accepted by all who had a hand in science policy. But when federal budgets became strained because of the Vietnam war, those holding federal positions took the position that there was no justification for a foreordained growth rate. Thus, in 1967, Donald F. Hornig declared, "We accept as the goal that America must be second to none in most of the significant fields of science. . . . What is *not* accepted is the notion that every part of science should grow at some automatic and predetermined rate, 15 percent per year or any other number, as a consequence." (Address to the American Physical Society, Washington, D.C., April 26, 1967.)

tionship, the advisory structure holds a key initiative in the decision-making process: it more or less identifies the problems, states the issues, frames the questions, and brings forth a good deal of the data, all of which are then employed in the clouded process of deciding who gets what from the United States Treasury. The Westheimer report was, in effect, an indictment of existing policy and practice in federal support of chemistry. And, since the White House science office had informally agreed to give serious consideration to COSPUP reports, the entire federal system for supporting research—from the White House down through all the agencies that award grants and contracts—was essentially placed on the defensive; it had to show either that Westheimer was in error, or if he was correct, just what was to be done about it.

Committees being the vehicles for doing business in the government of science, the White House science office responded to the report of the Westheimer committee by appointing an interagency committee to evaluate the data and recommendations. The conclusions of this committee were that chemistry was actually receiving more money than the Westheimer group had chosen to acknowledge; that it was difficult to justify a sharp jump in federal support for chemistry when every other discipline could legitimately claim that support had long lagged behind its crucial needs.* But, in the end, as one White House science aide put

* The committee also concluded that one reason chemistry had not fared especially well in obtaining NIH funds was that chemists sent relatively few grant applications to NIH. This finding raises the subtle and difficult problem of "proposal pressure." When a particular discipline, for whatever reason, concludes that a federal agency is especially interested in its field of research, the number of applications for support is likely to shoot up rapidly. Since the federal system is structured to respond to the initiative of the working scientist, an increase in applications is likely to be followed by an increase in grants. If, within an agency's various disciplinary subdivisions, funds are not available to meet the growth in applications, the inundated program director can argue that he should be allotted a greater proportion of the overall budget because he is unable, with his present budget, to support a reasonable proportion of the worthy applications he has received. This being the case, there is an incentive for scientists to deluge an agency with their applications, and program directors have been known to go out and drum up a flood of applications. But, since it is a lot of work to prepare a grant application, once the word gets around that funds are not forthcoming, scientists often lose interest and no longer even bother to seek funds from agencies with a reputation for being unresponsive to their discipline. Thus, while NIH contends that it is as interested in chemistry as it is in any other field, chemists do not think of NIH as their agency, there is little proposal pressure at NIH in the field of chemistry, and NIH can reasonably conclude that, in terms of granting what is sought, it is doing as well by chemistry as by any other field.

it, "On the whole we agree with Westheimer." A direct consequence of this was that NSF, whose financial decisions regarding chemistry had inspired the report, agreed that it would be desirable to make greater sums available for chemistry. But with the costs of the Vietnam war impinging on all federal activities, the Johnson administration—though treating research budgets relatively well—was not disposed to give NSF the sum that NSF officials felt could be well used, and Congress, following its traditional pattern of dealings with NSF, was not disposed to vote even that amount. Nevertheless, the Foundation strained as much as it could, and, from having expended $12 million for chemistry in fiscal 1965, it went on to obligate $18.8 million for 1966, $20.3 million for 1967, and $24.5 million for 1968. While other agencies did not respond with similar levels of generosity, it cannot be said that the chemists had done too badly. They had carefully stated their grievance to the government of science, and, in its own peculiar ways, the government had responded with at least partial relief. Perhaps even more important was the matter of image. Because of the budgetary situation that had grown out of the war, something of a moratorium had come into being as far as rapid growth was concerned for any field of basic science. But because of the report, the chemists now stood forth as the most deprived of the deprived, and it was their expectation that when the dikes once again opened, their needs would be attended to promptly.

Thus concludes our excursion with the chemists. To proceed further with our inquiry into the politics of pure science, we will now look at that great misadventure known as Project Mohole.

NOTES

1. *Chemistry: Opportunities and Needs,* National Academy of Sciences–National Research Council, 1965, p. 10.
2. Annual budgetary figures compiled from *The National Science Foundation, A General Review of Its First Fifteen Years,* Report of the Science Policy Research Division, Legislative Reference Service, of the Library of Congress to the Subcommittee on Science, Research, and Development, of the House Committee on Science and Astronautics, 1965, p. 32. Figures on allocations for chemistry supplied by the NSF Office of Information.
3. "Government Influence on the Conduct of Scientific Research," paper delivered at Conference on the Future of Science in the Liberal Arts College, Wheaton College, Norton, Mass., April 15, 1967.
4. *Ibid.*

5. "Doctoral Education in Chemistry. A Report of Current Needs and Problems," *Chemical and Engineering News*, Vol. 42 (May 4, 1964).
6. *House Independent Offices Appropriations, Hearings*, 1964, p. 326.
7. William D. Carey, "Research, Development and the Federal Budget," address to the 17th National Conference on the Administration of Research, Estes Park, Colorado, September 11, 1963.
8. *Chemistry: Opportunities and Needs*, p. 14.
9. *Basic Research and National Goals*, report to the Committee on Science and Astronautics, House of Representatives, by the National Academy of Sciences, 1965, p. 13.

IX
Mohole:
The Anatomy of a Fiasco

. . . the Mohole project is a very complicated situation. I have been
in some complicated situations in my life, but I have never been in one
that approached this for complexity.

> LELAND J. HAWORTH, Director, National Science
> Foundation, before the Subcommittee on Science,
> Research, and Development of the Committee
> on Science and Astronautics, 1963

Alas, pure science! NSF, Vannevar Bush's fragile flower, is right now
being wilted by bureaucratic contamination. . . .

> HERBERT SOLOW, "How NSF Got Lost in Mohole,"
> *Fortune*, May 1963

In the entire postwar partnership of science and government, no research enterprise has produced greater political embarrassment, anguish, or defeat for the scientific community than the grandiose and
lately departed project known as Mohole. As a scientific undertaking
Mohole was unique and meritorious, its object being to drill an un-
precedentedly deep hole into the ocean floor for the purpose of studying the composition of the earth's interior. Mohole never achieved this
object, but it did attain an assortment of distinctions, ranging from cost
estimates that rose from $5 million to $125 million, charges of political
favoritism in the award of contracts, open and at times virulent contention among the normally discreet statesmen of science, and finally, the
greatest distinction of all, namely, that of being the only basic research
project ever terminated by Act of Congress. As we approach the Mohole case, it should be noted that it can be studied for entertainment
value, of which it provides a great deal. It can be used for flogging contemporary science for worldliness and questionable values, and this has

been done, not without justification.* Or, as I propose to do, Mohole can be subjected to sober inquest to make use of the opportunity it affords for a unique view of the inner workings of the scientific community under conditions of stress.

In 1952, at the Office of Naval Research, in Washington, two staff geophysicists, Gordon Lill and Carl O. Alexis, were sorting applications for research support. Lill and Alexis soon noticed that existing research categories were inadequate for accommodating the diversity of the proposals, and, as a consequence, a good number had to be assigned to the "miscellaneous" pile. Whimsy being their mood, they then and there founded the American Miscellaneous Society—conveniently referred to as AMSOC—and in short order many outstanding figures of American geophysics proudly asserted their affiliation with this latest addition to the rolls of professional organizations. It is clear, however, that AMSOC was simply a humorous twist upon what has been referred to as the "invisible colleges" of scholarship—small, informal networks of people who work in the same field, and who, on the basis of personal or professional acquaintanceship, easily communicate about the stuff of their trade. AMSOC's *raison d'être* simply was the comic contrast that it provided to the traditional ways of established scientific societies. In fact, if it had been any less of an organization, it would have been nonexistent.

With a great sense of camaraderie and perhaps even a feeling of naughtiness for dispensing with the ponderous ways of the established professional organizations, AMSOC members ostentatiously disavowed membership rolls, bylaws, officers, elections, publications, and formal meetings. When they did concede to form, they were facetious all the way. Like all scientific societies, AMSOC had disciplinary subgroupings; however, in the case of AMSOC these were alleged to be in Etceterology, Phenomenology, Calamitology, Generalogy, and Triviology. AMSOC also maintained relations with the Committee for Cooperation with Visitors from Outer Space, as well as with the Society for Informing Animals of Their Taxonomic Position. Like all scientific societies, it awarded an annual prize for outstanding accomplishment. AMSOC's prize was the Albatross Award. A quorum of two sufficed

---

* See, for example, Jerome R. Ravetz, "The Mohole Scandal," *The Guardian,* October 27, 1964, which observes that "the norms of social behavior appropriate to politics, business enterprise, or even speculative technology are not those of science. When they are imported into science, they constitute corruption."

for AMSOC meetings, and these usually took place over drinks at the Cosmos Club. AMSOC performed no visible function beyond delighting its amorphous "membership," which tended to be concentrated in geophysics and earth sciences, the fields of its founders. Occasionally, to the great self-satisfaction of those associated with AMSOC, marvelously grotesque "scientific" propositions would emanate from its "meetings." At one of these, for example, detailed calculations were prepared to demonstrate the feasibility of solving Southern California's water problems by towing antarctic icebergs to Los Angeles and melting them into the water system. Socially, AMSOC was delightful; scientifically, it was inconsequential, except to the extent that its "members" shared common professional interests and, when together, were more likely to engage in shoptalk than in whimsy; politically, in the affairs of the scientific community, AMSOC neither sought to be an influence, nor, considering the stuffiness of the community, could this prankster organization reasonably aspire to influence.

Such were the first five years, which brings us to the spring of 1957. This was a time, it is important to note, when both the amount and rate of growth of federal support for many fields of science had become strikingly large. Physics was doing splendidly with the Atomic Energy Commission; the biomedical sciences were entering a period of extraordinary growth underwritten by the burgeoning budget of the National Institutes of Health; and space researchers, though later wishing they had thought bigger and moved faster, were laying the foundations for a great program of space exploration. However, largely left out of this affluence were the earth scientists, for whom, as was the case with the chemists, no exclusive source of support existed in the hierarchy of federal research agencies. The ancient United States Geological Survey, deeply embedded in the Department of the Interior, was the focal point of government interest in the earth sciences, but the Survey was small, conservative, and most damning of all, as far as academic science was concerned, did not give grants; its funds were almost entirely for the use of its own staff members. NSF and ONR put some money into the earth sciences, but in both these agencies, the earth sciences had to share the available wealth with dozens of other fields of research.

These being the economic facts, the spirit of chauvinism and evangelism dictated that the earth scientists must do something to improve their position. That something turned out to be Project Mohole, and, fortunately, there is a remarkably frank account of the moment of conception from one who was there, Harry H. Hess, chairman of the Princeton geology department. Speaking at an earth sciences symposium in 1965, when Mohole, having overcome much travail, was alive and apparently en route to success, Hess spelled out the following tale of genesis:

"The Mohole project started in March, 1957, at a National Science Foundation panel meeting where eight scientists were gathered to analyze projects submitted from earth scientists of the country. We had something like sixty projects to review in two days. A week's work was involved in reading these projects before we met. At the end of a two-day session we were rather tired and Walter Munk [a renowned oceanographer at the University of California Scripps Institution of Oceanography] mentioned that *none of these proposals was really fundamental to an understanding of the earth*, although many of them were very good. If someone wanted to study the clay minerals of such and such a formation . . . well, this was a good thing to do; *however, it would not solve a major problem about the earth*. . . . We had gone through sixty of these proposals, most of which we rated as being very good projects; they should be supported. Walter Munk commented that we should have projects in earth science—geology, geophysics, geochemistry—*which would arouse the imagination of the public, and which would attract more young men into our science*. We are very short of geophysicists, for example, and we were very short of oceanographers also at that time. It is necessary at times to have a really exciting project. . . . Walter Munk suggested that we drill a hole through the crust of the earth. I took him up and said let's do it; let's not drop it here, and we did go on." [1] *

Now, all of those present were well aware that a prodigious feat of engineering would be required to drill through the crust of the earth. Covering our planet like a bumpy orange skin, the crust stands in thicknesses of 3 to 40 miles; immediately below it lay the objective of the drilling venture, the mantle, a dense, rocky mass, some 1800 miles thick, which geophysicists wanted to sample as part of their efforts to answer multitudinous questions about the origins and history of the earth, the formation and drift of continents, the flow of heat from the earth's interior, and the formation of mineral resources.

---

* The italics are my own, for purposes of emphasizing two of the key motivations of Project Mohole: science and publicity. Munk confirms Hess's account as an accurate portrayal of the project's birth.

The progression from crust to mantle was believed to be indicated by a change in the velocity of seismic transmissions, a change which was first detected by a Yugoslav geologist, Andrija Mohorovičić, thus leading to the term Mohorovičić discontinuity, conveniently shortened to "Moho," and inevitably transformed into "Mohole" when the scheme to pierce the crust got underway. But in Project Mohole, the name was virtually the only thing that was to come easily. As if by a gross perversity of nature, the crust covering the mantle was so distributed that it was thinnest under the deep sea and thickest on dry land. At ocean depths of 14,000 feet near Puerto Rico, a survey later indicated, it might be possible to reach the mantle by drilling 18,000 feet into the bottom. On land, it would be necessary to drill at least 50,000 feet. The deepest hole yet drilled on land was 22,570 feet—which drilling engineers considered a stretching of the state of the art. At that depth, heat, corrosion, and the weight of nearly five miles of drill string severely strained existing technology. At sea, the deepest drillings had gone merely a few hundred feet. If, along with great new scientific understanding, the progenitors of Mohole sought "a really exciting project," something "which would arouse the imagination of the public," drilling to the mantle was undoubtedly the best they had to offer. The scientific motivations were valid and substantial—samples of the mantle could help answer many major questions in geophysics. But the circus aspect also was substantial. As Willard N. Bascom, an oceanographic engineer who became director of Project Mohole, later wrote, Munk's proposal to drill to the mantle evoked the observation, "This would be the perfect antianalogue of a space probe. Think of the attention it would attract to the earth sciences." [2]

And this brings us back to the American Miscellaneous Society. How was the drilling venture to proceed? Turn it over to AMSOC, Hess suggested. The following month, at a "wine breakfast" at Munk's lovely home on a coastal hillside in La Jolla, California, AMSOC made its first compact with conformity: it chose a chairman, cofounder Gordon Lill, who was elected by acclamation, and it formed a Deep Drilling Committee to accept the challenge of drilling through the earth's crust. Nominated to serve on the committee were Hess and Munk; Harry Ladd and Joshua Tracey, both of the United States Geological Survey, and Roger Revelle, director of the Scripps Institution of Oceanography. As a group, they were well connected with the substantive frontiers of their profession as well as with the political, and especially military, underpinnings that supported their work. Revelle, for example, had served on the staff of the Office of Naval Research in 1946 and 1947, and was a consultant to ONR; Hess and Munk were both members of the Academy, and the former, a naval reserve captain who had held a seagoing command, was also chairman of the Academy's earth sciences division. Ladd had served on the scientific staff of one of the early postwar atomic tests in the Pacific.

One week after the La Jolla proceedings, the members of the newly formed committee were filing into the Cosmos Club for their first meeting, when they encountered there in the lobby one of the senior eminences of their profession, Maurice Ewing, director of Columbia University's Lamont Geological Observatory. Ewing recalls that he was passing the time while waiting for an engagement unrelated to the affairs of AMSOC. His colleagues told him of the venture and urged him to come along and join AMSOC. When he protested that his professional interest was the sediments on the ocean bottom, not the deep underlying rock, he was told, "Maurice, you're thinking too small." He joined AMSOC, later resigned in disgust, but subsequently was persuaded to return.

If Ewing was thinking small, he definitely was out of harmony with his colleagues, for AMSOC was swept up with the excitement of the Mohole venture. At that meeting in the Cosmos Club, it was decided to ask NSF for $30,000 for a feasibility study. However, under the cautious and conservative hand of Waterman, the Foundation was not disposed to give funds to so ephemeral an organization as AMSOC. Taking this rebuff in stride, the once-freewheeling scoffers at institutionalism cheerfully withdrew their applications and took steps to bring their organizational facade into conformity with the preferences of the financial source they wished to tap. (Applicants are usually offered the option of withdrawal to avoid the ignominy of rejection.) The first step was for Hess, of AMSOC and the Academy, to ask the Academy if it would be willing to accept a grant from NSF to study the feasibility of Mohole. If there was any question about the desirability of such a study, Hess could point to a series of resolutions that had been adopted in recent months by the most prestigious professional societies in the earth sciences, for shortly after the meeting at Munk's home and at the Cosmos Club, the International Union of Geodesy and Geophysics had gone on record as urging "the nations of the world and especially those experienced in deep drilling to study the feasibility and cost of an attempt to drill to the Mohorovičić discontinuity at a place where it approaches the earth." [3] The authors of this resolution were Hess, Revelle, and a British colleague. Similar resolutions were also adopted by the International Association of Physical Oceanography and the International Association of Seismology and Physics of the Earth's Interior. And, as if the gods of geophysics had decided to bestow their greatest blessing upon the venture, a Russian geophysicist rose at one of these meetings and proclaimed, "We already have the equipment to drill such a hole; we are now looking for the place." [4] Ten years later, the Russians possibly were still looking for the place, for they offered neither evidence nor claims that they had proceeded with the project. But East-West competition was money in the bank and AMSOC knew it.*

The Academy was indeed willing to accept NSF funds, and on April 2, 1958, an application for $30,000 went out to NSF from the National Academy of Sciences–National Research Council, Division of Earth Sciences, Harry Hess, chairman. Four days later, the Academy took the unprecedented step of draping its prestigious mantle over an outside committee. Since NSF would not stake a loose-floating AMSOC committee, the Academy had agreed to give it a home. Tradition dictated against such a step, but AMSOC, now grown to nine members, included five members of the Academy; Hess, of course, was an officer of the Academy, and as a consequence, goodwill and personal confidence overrode tradition. On April 8, 1958, just a bit over a year since Munk had proposed piercing the earth's crust, the AMSOC enterprise was formally constituted as the AMSOC Committee of the Division of Earth Sciences of NAS-NRC. Since the frivolity inherent in the title

* In June 1966, following conversations in Moscow with high-level Soviet science administrators, Edward Wenk, Jr., chief of the Science Policy Division of the Library of Congress, Legislative Reference Service, concluded: "1. The Soviet Union is currently engaged in an experimental program of superdeep geological drilling. 2. There are *no plans* [original italics] to drill explicitly to the Mohorovičić discontinuity in the near future." ("Soviet Plans for Deep Drilling Related to Project Mohole," LRS, unpublished paper.) Wenk's findings were in harmony with an earlier LRS study, which concluded that "all evidence points to the fact that the USSR has not undertaken deep drilling for propaganda purposes, or to counter Project Mohole." All along there had never been anything but the flimsiest evidence of Soviet interest in a Mohole project, but this did not prevent the proponents of Mohole from projecting the impression of a frantic race with the Russians.

"Miscellaneous Society" was not consonant with the somber mien of the Academy, it was decided that henceforth AMSOC would be a word and not an abbreviation.

With the Academy affiliation there also came a chance occurrence that was to figure large in the affairs of Mohole. Next door to Hess's office in the Academy building worked a restless young engineer named Willard Bascom. Bascom had studied engineering and geology at the Colorado School of Mines, left without a degree, and then had gone on to a variety of jobs. During World War II, he was a mining engineer in Colorado, Idaho, Arizona, and New York. In the years after the war, he was a research engineer in oceanography at the University of California, lectured on oceanography at the Navy's Postgraduate School, and took part in monitoring the effects of nuclear weapons testing in the Pacific. In 1954, Bascom joined the staff of the Academy to handle a variety of jobs. On a staff that was not especially loaded with ambitious performers, Bascom quickly stood out. In 1957, he served as United States delegate to the International Geophysical Year Conference on Oceanography, in Sweden; he also became executive secretary of the NAS-NRC committees on meteorology and maritime research. With the excitement of Mohole spilling out into the corridors of the Academy, Bascom inevitably became interested in the project, and not long afterwards was asked by Hess to become AMSOC's executive secretary and to organize and conduct studies on how the project was to be carried out. Normally, Academy staff men are submerged and cautious behind the commuting luminaries who reap the glory, but the degreeless engineer Bascom was not disposed to cautious anonymity, or any great reverence for the prestigious statesmen of science. An energetic and extraordinarily inventive engineer, he interpreted Hess's mandate generously, and began efforts to make Project Mohole a reality.

Now housed in the Academy, AMSOC looked properly respectable, and, in response to its application for $30,000 to get the project underway, NSF ladled out half the requested amount, which was characteristic for the financially pressed foundation. Mohole was now a going concern—but going toward what? Since this question was to lacerate the normally harmonious relationship between NSF and the Academy, and eventually tear apart the project, it is worth noting with some care just what it was that AMSOC proposed when it sought money from NSF.

The application,[5] bearing the names of all the AMSOC members, opened with the statement that funds were requested "for support of the study of the feasibility of drilling *a hole to the Moho discontinuity* [italics supplied]." It then went on to list the scientific benefits that might ensue "If an authentic sample of the material below the discontinuity were obtained." It made reference to various measurements that could be made in *"the hole."* But it also stated the importance of obtaining samples of the sedimentary materials that lie above the mantle. Was this simply a proposal to drill one hole down to the mantle, pull out samples along the way, pack up, and terminate the project? Considering the cost of the drilling equipment, it would be extravagant to design and build it for the purpose of drilling one hole, as extravagant as it would be to design and build a telescope for simply one look at the heavens. But the application clearly suggested, and NSF then so interpreted it, that AMSOC was thinking of a one-shot affair.

In any case, with the NSF money in hand for a feasibility study, Mohole was officially under way, administered through the following organizational patchwork: AMSOC was responsible for supplying scientific guidance, but since AMSOC was a part-time organization, and its executive secretary, Bascom, was full-time, the energetic Bascom felt few restraints in running the show. Meanwhile, the Academy provided an institutional base and NSF paid the bills. For both NSF and the Academy this arrangement was a departure from their traditional forms of operation. In its role of "banker" for basic research, the Foundation normally gave grants only to educational organizations actively involved in the conduct of scientific research. Its recipient in this case, the Academy, had no educational function whatsoever; but more important, as a matter of practice, the Academy normally eschewed any active role in research, on the premise that, as statutory scientific adviser to the federal government, it should not be operationally involved in any of the matters on which its counsel might be sought. Eventually, the two institutions were painfully racked by these arrangements, but in the early days of Mohole, goodwill, excitement, and hope obscured seemingly unimportant departures from administrative form. Who was in command of the project and what was its objective? Like a boatload of efficient and harmonious oarsmen, everyone was pulling together; the issues of leadership and direction did not arise.

At several meetings in the fall of 1958, AMSOC reached out further into the geophysical community to enlarge its membership, and three panels were established: Hess headed the panel responsible for locating a drilling site; Ladd was named to head the panel on scientific objectives and measurements, and William B. Heroy, Sr., a veteran petroleum geologist, headed the drilling panel. As for the cost of the project, everyone concerned agreed that it would be high in relation to the funds normally available for the earth sciences, but compared to the gusher of funds that Sputnik had loosed into the space establishment, the amounts were trifling. In an article in *Science* co-authored by Arthur Maxwell, of the ONR staff, and AMSOC chairman Gordon Lill, it was confidently stated: "The Mohorovičić Discontinuity project probably can be accomplished for $5 million. Earlier and larger estimates were out of bounds. Five million dollars is a lot of money, but compared with the many millions of dollars that are being spent on moon rocketry and the billions spent on atom bombs, this is not an overly ambitious scientific venture." To which they added:

"The American Miscellaneous Society, with its flair for seeing the lighter side of serious problems, likes to quote the following proverbs when discussing the 'Moho': (i) 'When going ahead in space, it is also important to go back in time.' (ii) 'The ocean's bottom is at least as important to us as the moon's behind!' " [6]

At about the same time, to help explain the project to an increasingly interested public, Bascom spelled out its past and speculated on its future in an article in *Scientific American*.[7] The article, though not an official statement of the AMSOC committee, undoubtedly had its endorsement and, at that point, was the most definitive statement of what AMSOC—the originator of Project Mohole—thought it had in mind when it sought support from NSF. But conceptually, it introduced some important new elements into the project—particularly the idea that Mohole was *not* a one-shot affair. Wrote Bascom: "The principal objectives of drilling to the mantle are . . . to obtain samples of the various rocks of the mantle and the deep crust. . . . Although reaching the mantle is the ultimate objective of the Mohole project *an intermediate step* is likely to yield *equally valuable and interesting information.* . . . *No one site or hole* will satisfy the requirements of the Mohole project [italics supplied]."

Bascom then went on to suggest a four-step program, starting with a 260-foot former Navy freight barge, the *CUSS I,* that had been converted into a drilling barge by the four oil companies whose initial letters comprised its name, Continental, Union, Shell, and Superior. First,

wrote Bascom, studies would have to be made to improve upon the
best existing drilling equipment and techniques; second, modifications
would have to undergo trials at sea; third, "when the results of the first
tests are in, modifications will have to be made in the drilling methods;
*this in turn may require a practical readjustment of scientific require-
ments.*" Finally, "the hole to the mantle will be drilled. It will be very
difficult; there are no illusions about that. But it will remain the ultimate
objective of the Moho project." In the course of the article, Bascom
incidentally noted the Soviet Academy of Sciences had appointed a
committee similar to AMSOC. "Perhaps there will be a race to the
mantle," he observed.

Now the administrative pace of Mohole began to accelerate. The
National Science Foundation awarded grants totaling $80,700 for
drilling site surveys to be conducted by the Woods Hole Oceanographic
Institution and the Lamont Geological Observatory. In neither case did
the Foundation consider it necessary to scrutinize the venture it was
funding, and, for that matter, why should it have? The scientists associated with Mohole were the outstanding figures of geophysics, and,
in some instances, were among the government's principal advisers on
geophysics and earth sciences. Hess and Heroy, for example, were on
NSF's advisory panel for earth sciences. Anyone knowledgeable in these
fields would agree that sampling the mantle was an important scientific
objective, one that would greatly contribute to the solution of many
questions of importance to geophysics. As for the question of what
priority the Mohole project should occupy in the general scheme of
federal support of research—that simply never arose. The ideology of
basic science held that all unanswered scientific questions were equal;
the only inequality lay in the ability of the various researchers to tap
into the federal treasury. And, in the case of the Mohole, those interested in reaching the mantle were eminently able to do so because of
the esteem with which they were regarded at NSF.

In early June, within a matter of months after Lill and Maxwell had
published the assertion that the "project can be accomplished for $5
million," AMSOC's executive committee reconvened at La Jolla and
decided upon a tentative budget of $14 million "for the entire project." [8]
What conceivable explanation is there for the great disparity between
the two figures? Part of the answer is that these were scientists, not

engineers; they were neither trained for, nor accustomed to dealing with, the complexities of estimating the design, construction, and operating costs of a major engineering project. But what is perhaps more important, the atmosphere pervading big science and technology was such that there was no pressure to be mindful of economics. As one White House science adviser later explained his own lack of close attention to Mohole: "I took a quick look at the project and decided, 'Why not. It's only going to run about the cost of one space shot.'" At that same meeting where the $14 million figure was introduced, it was decided that AMSOC should actually run the project—with Bascom elevated from staff obscurity to the position of project director—and not merely serve on the sidelines as a scientific adviser. The governing board of the Academy was amenable to this, though there is nothing to suggest that that part-time body ever examined closely the implications of putting the Academy into the business of running a pioneering deep-sea drilling program. Meanwhile, NSF awarded another $80,500 to pay the salary and expenses of the now-blossoming AMSOC staff. A few months later, as plans began to jell, AMSOC, upon application to NSF, was assured of up to $1,250,000 to get under way with what was essentially Bascom's four-stage drilling program. And not long afterwards, the Washington Pick and Hammer Club, composed almost wholly of geologists and geophysicists, put on its annual show. The title that year was *Mo-Ho-Ho and a Barrel of Funds*.

Now, with speed and success that obscured some of the difficulties of drilling all the way to the mantle, AMSOC's professional staff carried out Mohole's first phase, a series of test drillings in the vicinity of La Jolla and Guadalupe Island. The drilling vessel was the one suggested by Bascom, the converted Navy barge *CUSS I*, operated under contract by its owners, the Global Marine Exploration Company, but modified, equipped, and run under Bascom's direction. Early in 1961, the *CUSS I* set out from San Diego to test the feasibility of drilling in very deep water—ten times deeper than any in which oil drilling attempts had been successful. Of critical importance to the operation was the positioning system devised by Bascom for holding the vessel relatively motionless on the open sea while the drilling bit chewed into the ocean bottom. Since the depths at which the Mohole could be most easily reached were too deep for anchoring a heavy vessel, Bascom sought the required stability with a dynamic positioning system that employed four centrally controlled 200-horsepower outboard motors located around the hull of the *CUSS I*. Surrounding the vessel was a ring of buoys to provide reference points. As wind or current worked to push the *CUSS I* away from its station directly above the drilling site, the appropriate motor, operated by a single control, would push in the opposite direction. The system, which had aroused considerable skepticism when Bascom first proposed it, was able to hold the *CUSS I* virtually motionless in a twenty-knot wind.

The *CUSS*'s performance was spectacular, and culminated in a record-smashing boring of 601 feet into the ocean bottom, in water about two miles deep. And Bascom's AMSOC crew completed the operation within a matter of weeks and within the budget they had stated beforehand. Once again, all this took place in a period of extreme goodwill, when little concern was paid to organizational detail. Bascom, though employed by the Academy as director of the project, was designated a technical representative by NSF. As he later explained before a congressional committee, "This method was an administrative makeshift, but it worked, primarily because all of us were anxious to make the tests a success." [9]

Successful as the first phase was, it had literally only scratched the soft bottom of the ocean, for to get to the mantle, it would be necessary to work in some three miles of water and drill about 15,000 feet into the floor—most of the way through rock so dense that at least two years of nonstop drilling would be required. No equipment capable of this feat was in existence. Thoughts now turned to Phase Two of the project, and it was at this point that strains began to be felt in the administrative patchwork that had theretofore served so well. It was the spring of 1961, nine years after AMSOC's birth, four years after Munk had first proposed piercing the earth's crust, and a period of stocktaking was at hand. Bascom and his AMSOC staff triumphantly wrote up the *CUSS I* experience, *Experimental Drilling in Deep Water,* which was issued as an Academy document.[10] Included was a congratulatory message from President John F. Kennedy, addressed to Academy President Bronk and NSF Director Waterman. Kennedy extended his congratulations to all associated with Project Mohole "and especially to all those on board the *CUSS I* and attendant vessels who have combined their talents and energies to achieve this major success." Amidst success, there was no sign that events were so proceeding that the much-lauded Bascom group would soon be cast out of the project.

For the conservative and tradition-minded Academy, AMSOC's presence as a committee of the Division of Earth Sciences was disconcerting. Unlike its Soviet counterpart, which was deeply involved in the direct administration of research, the Academy had so conducted its affairs as to stand in pristine aloofness from any matters on which its counsel might be sought. Now, with Mohole's Phase One successfully completed, the Academy's governing board decided it was time to get out. On June 13, 1961, Academy officials told AMSOC that the Academy, while "desiring to advance the Mohole project in every way, urges that means be found that will not involve the Academy-Research Council in the actual operation of the project. . . ." [11] Thus, the Academy wished to disengage itself. But, as it turned out, so did AMSOC. The day after the pronouncement from the Academy AMSOC Chairman Lill informed the Academy of a decision that had been reached by the committee. Writing to Academy President Bronk, Lill stated: "It appeared to the Committee that the administrative demands of continuing work at sea, and the deliberate pace of the Committee activities do not mix well, and that we must reexamine our position. We decided that the AMSOC committee should be [sic] in the future concern itself with matters of scientific policy, engineering review, and budget. . . . We consider that we are responsible to both the National Academy of Sciences–National Research Council and the National Science Foundation. The dual responsibility arises because of our origin and existence in NAS-NRC, and because of our financial support from NSF. In this relationship we may properly act as the representatives of NAS-NRC in its role as adviser to the National Science Foundation for drilling to the mantle." [12] If AMSOC was to assume a purely advisory role, who would actually run the project? AMSOC's recommendation was that Phase Two of Mohole should be contracted out to an industrial or academic organization, and it offered a list of eleven possible candidates—nine oil or oil drilling companies, an aerospace firm, and the Scripps Institution of Oceanography—from which it said it had "received, or learned of, expressions of interest." As for Bascom's group, AMSOC stated that "under our auspices for purposes of establishing Project Mohole, there has been assembled a technical staff that desires as a body to continue with the project. The AMSOC Committee desires to retain the staff intact on the project. We recommend, therefore, as a part of the terms of the prime contract, that the contractor must agree to make our staff the nucleus of his endeavor by absorbing them into his organization." The recommendation ran counter to established patterns of organizational administration. Why should a contractor accept outside dictation on who is to staff his organization? AMSOC apparently recognized the presumptuousness of its proposal, for it noted, "These terms may have inherent difficulties, but we may not overlook our responsibility to the staff, at least some of whom may well make ocean basin exploration their life profession." And, in the expectation that goodwill, as in the past, would override all difficulties, AMSOC added, "If difficulties arise, then the AMSOC Committee or its executive committee will promptly meet with its staff to seek appropriate resolution."

Having proposed an administrative setup largely based on wishful thinking, AMSOC proceeded to set forth a program so marvelously kaleidoscopic that it easily could mean whatever any partisan chose to make it mean. The only certainty spelled out in the budget was that for fiscal 1962, which began in sixteen days, a minimum of $4,250,000 would be required. How this was arrived at is clear: AMSOC simply threw together all the figures it could, reasonably or otherwise, include, such as $50,000 for "turbodrill studies" in France; $10,000 for "drilling tools"; sixty days at sea, at $3500 per day, for a ship, yet to be designed, built or tested, whose construction and equipping was budgeted for some $2 million. Also included was an $80,000 item to carry the AMSOC staff for ninety days until it was transferred to the proposed contractor, after which the support of the staff would be met out of a $600,000 item for the salary and administrative costs of the contractor.

Toward what objective was this wealth to be directed? On the basis of AMSOC's own pronouncement, unanimously adopted, NSF could reasonably conclude that a hole to the mantle was the objective—and nothing else. "We are agreed," AMSOC reported, "that the major scientific objective of Project Mohole is to drill to the earth's mantle through a deep ocean basin. . . . Also exciting, and of prime scientific importance, is the fact that we now have a new tool, the floating drilling vessel, with which to explore thoroughly the sediments and upper crustal layers of the ocean basins. We find, however, that the major objective of the Committee will entail work enough, and that *we must recommend this possible exploration program for separate scientific and financial consideration* [italics supplied]." But at the same time, AMSOC acknowledged uncertainty on the question of whether sufficient data had been accumulated with *CUSS I* to design a vessel that could drill to the mantle. Bascom, who had more experience in drilling on the high seas than any member of the AMSOC committee, insisted that it would be necessary to build and experiment with an "intermediate" ship to obtain the data for the "ultimate" ship. But Bascom was a staff employee of Project Mohole, not a policy maker. Furthermore, the committee that employed him was disengaging itself from operational involvement in the project, and was putting him and his staff up for employment by a still unselected contractor. On the other hand, AMSOC's own drilling techniques panel had concluded that "the ultimate ship should be developed immediately rather than an intermediate ship," [13] citing as one reason an expression of urgency from the President's Science Advisory Committee (PSAC). Why this top-level advisory body should have felt urgency in this matter is not clear, but technological spectaculars —such as the manned lunar program suddenly announced by Kennedy a few months earlier—had become a highly effective technique for providing a financial uplift for all fields of science and technology, and possibly PSAC regarded Mohole as a master key for opening the treasury to all the earth sciences. In any case, both the White House and AMSOC's own drilling panel said the "ultimate" ship should be the next step. But not long before, Roger Revelle, one of the charter members of AMSOC, had gone before a congressional committee to testify that he would "divide the drilling into parts." He would "drill a lot of holes through the sediments . . . several hundred holes . . ." And he would drill the Mohole.

Against the background of divided opinion, AMSOC fudged the "intermediate-ultimate" question. It strongly recommended that an intermediate drilling *program* be undertaken in the coming fiscal year, and it stated that it based the $4,250,000 budget on the assumption that an intermediate *ship* would be employed, but then, in open disagreement with its hired staff and in apparent deference to the views of the drilling panel, it went on to state that it might be possible to skip building an intermediate vessel and go directly to the construction of a deep-drilling ship. How was this question to be resolved? AMSOC's answer was that the decision is "contingent upon the findings of the Drilling Techniques Panel working jointly with the AMSOC staff, and eventually with the prime contractor." Thus confusion and dissension, already well sprouted, were coming into bloom.

Ten days after AMSOC had formulated these recommendations and statements of intention, Academy President Bronk forwarded them to NSF with an accompanying note of approval. "In connection with the proposed transfer of the staff," he wrote, "I am glad to record my own admiration, and that of our Division of Earth Sciences and our Governing Board, for the exceptional performance of Mr. Willard N. Bascom and the staff members. . . . In our estimation, this group has been chiefly responsible for the successful carrying out of an undertaking that represents not only a scientific advance of unusual significance but also a distinguished engineering achievement and a major extrapolation of previous practice and experience." [14] At this point, as the summer of 1961 approached, all the signs were bright. But, in fact, Project Mohole was directly en route to political disaster.

Though NSF, with some minor, short-term assistance from ONR, was paying all the bills for Mohole, the fast-moving and ambitious project was essentially a sideshow for the management of the Foundation until 1962; then, NSF set up its own Mohole Committee, consisting of three NSF executives. But in the meantime, NSF ladled out the money and promptly accepted AMSOC's recommendation that a contractor be hired to handle the next phase. On July 27, 1961, in response to a public announcement, some two hundred persons, representing eighty-four organizations interested in the project, assembled in Washington to be briefed by NSF officials. (Since the preparation of a bid for a large scale engineering project is frequently quite costly, such briefings are often held to guide qualified firms and spare the time and effort of those that, for whatever reason, are not equipped to compete.) Now, what was it that NSF wanted the contractor to do? The announcement of the briefing clearly indicated that NSF realized that the project would have to feel its way, but it also indicated that NSF still conceived of Mohole as a single penetration of the earth's crust. Stated the announcement: "The Mohole project will include: (1) The conduct of deep ocean surveys, (2) the design and construction of deep drilling equipment, and (3) the drilling of a series of holes in the deep ocean floor, *one of which* [italics supplied] will completely penetrate the earth's crust." [15]

As for the "prerequisites" and "criteria" for selection, these were a mélange that could justify any decision. The contractor, NSF stated, should "have the necessary experience, organization, technical qualifications, skills, and facilities, *or the ability to obtain them* [italics supplied]." The contractor must also have "interest and enthusiasm in undertaking Mohole project." And, of course, as specified by AMSOC, there was a stipulation that consideration would be given to "whether the AMSOC staff [that is, Bascom and his associates] may be readily incorporated in the company for the duration of this job as part of a separate group entirely devoted to this work. . . ." Persons attending the briefing came away with the impression that NSF had indicated a preference to have the job done at cost, without any fee for the contractor.[16] However, the invitation to bid did direct applicants to specify the "rate of fee that would be charged prime contractor effort, overhead, and general administrative rates that would be charged to this project during the next twelve months." As is fairly common, NSF set forth a point system for evaluating bids, but clearly it was a point system that could be employed for justifying any decision. The greatest weight, 39 percent of the total, would be assigned to "management and policy considerations"—which were defined to include "objectives and attitude of organization in submitting this proposal with regard to the following: profit, national interest; publicity, to gain technical know-how advantage over usual competitors; and basic interest in research in the geophysical sciences." By the September 11 deadline, bids responsive to NSF's objectives had been submitted by eleven individual or combined organizations:

Socony Mobil, with General Motors, Texas Instruments, and Standard Oil of California

Global Marine Exploration, Aerojet-General, and Shell Oil, in a joint venture

Zapata Off-Shore, with the Electric Boat Division of General Dynamics and Dresser Industries (later, Continental Oil joined this team)

General Electric, with Kerr-McGee Oil, Youngstown Sheet & Tube, Petty Geophysical Engineering Co., Waldemar S. Nelson & Co., and Lamont Geological Observatory of Columbia University

Brown & Root

Westinghouse Air Brake's Melpar, with Bethlehem Steel, Geonautics, Inc., and four drilling companies

Litton Systems

System Development, a nonprofit corporation, with Offshore Co., a drilling company

National Engineering Science Co.

The University of California's Scripps Institution of Oceanography

Battelle Memorial Institute (nonprofit), with Scripps, Texas Instruments, Cornell Aeronautical Laboratory, and Ocean Drilling & Exploration

From this point on, NSF was to find itself in the most serious political trouble of its decade-long existence.

To evaluate the bids, Waterman established a preliminary evaluation panel of six NSF staff executives. The panel's conclusion was that the Socony Mobil group, with a score of 936 out of a possible 1000 points, was "in a class by itself." [17] The bid from the Global Marine group was rated a strong second, with 902 points. In fifth place, with 801 points, was an interesting latecomer to the bidding, Brown & Root, of Houston, Texas, a $250-million-a-year engineering and construction firm, which had not even attended the July briefing session. Brown & Root's bid had not arrived until just a week before the deadline; in contrast to Socony Mobil, which spent some $150,000 in working up original designs for solving Mohole's technical problems,[18] Brown & Root submitted a proposal that consisted of virtually nothing but promotional material for standard equipment and techniques. The preliminary evaluation panel saw no distinction among the firms that fell below Socony Mobil and Global Marine.

The next round of review was conducted by a panel of four senior NSF officials. This panel, too, concluded that Socony Mobil had submitted the most outstanding bid. There then followed a decision to cut the competition down to the top five and hold detailed reviews with each organization. Out of this procedure the Global Marine group came out with a slight edge over Socony Mobil, 968 to 964; in third place was Brown & Root, with 899. Members of the panels, after visiting the home offices of the three organizations, concluded they were all "competent to effectively complete the Mohole Project." [19] Perhaps the most striking difference was that both Global Marine and Socony Mobil told the Foundation that they would not charge a fee or overhead for managing the project; Brown & Root, on the other hand, had included in its bid provisions for overhead charges of 3 to 5 percent on all costs, plus a management fee of $1.8 million. The final decision now lay with

NSF Director Waterman. On February 28, 1962, he announced the selection of Brown & Root.

In awarding the contract to the Houston firm, NSF obtained the services of an engineering behemoth that had compiled an extraordinary construction record, one that included hundreds of combat vessels during World War II, 240 offshore drilling platforms for oil companies, and the twenty-four-mile bridge over Lake Pontchartrain, near New Orleans. Among Brown & Root's other distinctions, however, was its close proximity to the congressional district of Congressman Albert Thomas, who for a decade, without any notable generosity, had reigned over the NSF budget in the House. George Brown, president of the firm, was an intimate political ally of not only Thomas, but also of Vice President Lyndon B. Johnson.* Shortly after Brown & Root's selection, Waterman and other officials of the Foundation made their annual appearance before Thomas's subcommittee. The congressman was in an unusually affable mood. Thomas described the proposed NSF budget as "a work of art," and when the $5 million item for Mohole came up, his only part in the colloquy was to assist the NSF officials in answering other members of the committee. When one member, for example, protested the estimated expense—which NSF officials then put at $35 to $50 million—Thomas benignly interjected with, "The more important question is: What do you hope to gain by this experience?" [20]

---

* There have been allegations but never any concrete evidence of Lyndon Johnson's involvement in the award of the Mohole contract. There is no doubt, however, as to the existence of a long-standing and close relationship between Johnson and the brothers George and Herman Brown. Evans and Novak, *Lyndon B. Johnson: The Exercise of Power* (New York: New American Library, 1966), relates that Johnson-Brown dealings go back at least to the late 1930's, when Johnson helped the pair get a contract for construction of a federally financed dam (pp. 8–9). In 1947, the Browns' Texas Eastern Co. received government approval to purchase two federally owned pipelines for $143 million. "Now powers in the oil and gas industry, and already linked to Johnson, they quickly plunged ever deeper into Democratic politics—and became Lyndon Johnson's personal vehicle for a lateral movement into the center of the new oil power" (p. 18). The heart attack that Johnson suffered in 1955 occurred while he was a weekend guest at George Brown's Virginia estate. As for Thomas, though he was not widely known, he was properly regarded, by those familiar with the workings of government, as one of the most powerful men in Washington. His chairmanship of the House Independent Offices Appropriations Subcommittee afforded him virtually complete control over the budgets of 22 federal agencies, including NASA, the Veterans Administration, the Civil Service Commission, the Federal Communications Commission, and the Federal Aviation Agency. In 1963, his subcommittee reported out appropriations totaling $13.1 billion.

The protest was made by Congressman Joe L. Evins, of Tennessee, who was to succeed to the chairmanship upon Thomas's death in 1966—and whose first major act as chairman was to kill Project Mohole.

Waterman's selection of Brown & Root resulted in numerous allegations that political influence had determined the award of the contract. But if, in fact, Thomas did pressure Waterman into awarding the contract to his constituent firm, or, if Waterman did so with the intention of wooing the petulant Texas congressman, no direct evidence was left behind. In any case, though Brown & Root emerged with the contract after an extraordinarily circuitous selection process, Waterman adamantly insists that he never discussed the contract with Congressman Thomas prior to making the selection. As for the decision to choose a general engineering and contracting firm over two oil companies, Waterman says that both he and his staff members were concerned that the latter might be more interested in oil than in the purely scientific objectives of the project. On the matter of Brown & Root being the only firm to seek a management fee—Waterman explains that he felt it indicated a more "businesslike" attitude. At the request of Senator Thomas Kuchel, the General Accounting Office, which is Congress's fiscal investigatory arm, examined the contract award. It reported, "We are unable to conclude that the award to Brown & Root was not in the public interest." However, whether or not the selection was in the public interest, its appearances were so suspect that, in addition to the growing confusion over objectives, Mohole now bore the stigma of being involved in a questionable political deal.

While the contract selection process was bestowing a detrimental political image upon Mohole, there were developments within AMSOC that were to bring that body into open conflict with the Foundation. First of all, AMSOC Chairman Lill had left the Office of Naval Research to work for the Lockheed Aircraft Corporation. When that company thought it might enter a bid on the Mohole project, Lill resigned from AMSOC to avoid any appearance of conflict of interest. In the same period AMSOC also received resignations from a number of other members associated with firms interested in the contract. In one sense it would seem quite desirable to award the contract to a firm that had the services of someone intimately familiar with the project, but conflict of interest is a politically explosive issue and its avoidance was deemed absolutely necessary—at whatever price in the best use of skilled man-power. Shortly after Lill's resignation, an invitation to accept membership on AMSOC was accepted by Hollis D. Hedberg, a vice president of Gulf Oil and part-time professor of geology at Princeton University. Hedberg, unlike a number of other AMSOC members, vigorously devoted himself to the affairs of the committee, and two months after his appointment, succeeded to the chairmanship left vacant by Lill's resignation.

While Waterman was signing up Brown & Root to build a ship for "the drilling of a series of holes in the ocean floor, *one of which* will ultimately penetrate the earth's crust," Hedberg was educating himself in the muddled affairs of AMSOC. After several meetings in January with his colleagues on the committee, the new chairman became convinced of the folly of a one-hole Mohole project, and thereupon energetically set out to dispel the idea that Mohole was simply a design to penetrate the earth's crust. Just a few months prior to Hedberg's appointment, AMSOC had declared that the "major scientific objective of Project Mohole is to drill to the earth's mantle," and that exploration of the sediments and upper crustal layers should receive "separate scientific and financial consideration." But now Hedberg, in a long, detailed memorandum addressed to members of the AMSOC committee, stated, "I am strongly inclined to favor the recognition of a broad-gauge general objective for this project, leaving the way open for accomplishment to come along any or all of several specific lines. It seems to me that the overall ultimate purpose of the project can be simply stated as *to contribute to the determination of the nature and characteristics of the as yet unknown portions of the Earth's crust and mantle* [original italics] . . . I should perhaps reiterate that the drilling to the upper mantle—the Mohole—should be the central theme of the project because it is the ultimate goal. . . . However, with respect to the value of scientific information to be obtained, who is to say where the maximum value will lie in this realm of the unknown? Certainly it is easy to see many objectives in oceanic deep drilling in addition to the penetration of the Moho, which are deserving of more than incidental attention. Many of these may have a scientific value equally as great as the Moho and many may be reached more easily and at much lesser depth than the Moho."

Hedberg then went on to break completely with the original concept of Mohole as an undertaking focused on penetrating the mantle: "The project which AMSOC has launched should in no way be considered merely a stunt in deep drilling; it is a most serious search for fundamental information concerning the rock character and the rock record of the earth on which we live. And the scope of the project should be such as to take advantage of opportunities for contribution to this geological end *wherever they may be found—water or land, deep or shallow* [italics supplied]."

With the project conceived in this fashion, Hedberg continued, there should be no attempt to proceed directly to the construction of the "ultimate" ship. The program should proceed, he stated, with the construction of an intermediate ship to conduct an intermediate drilling program, followed by construction of the ultimate ship. ". . . The advantages of two ships are so compelling," he argued, "as to seem to demand an effort in this direction." The smaller ship could be drilling at sea while the larger vessel was under construction. Further, he pointed out, "Even the so-called prestige angle calls for this procedure since the blow of possible failure in one operation can always then be cushioned by the continuing success in another." [21]

Against this statement of Hedberg's views, the critical point to keep in mind is that Waterman was letting a contract to build a ship to drill one hole to the mantle. As for Bascom's group, which was now cut loose from the Academy, it incorporated itself as Ocean Science and Engineering (OSE) and went into the business of oceanographic consulting. Brown & Root fulfilled the contract obligation by hiring OSE as consultants for Project Mohole. But the irascible and capable Bascom had no stomach for the strangely chosen recipients of the project that he, more than anyone else, had successfully led. Within two months, relations between OSE and Brown & Root had so deteriorated that Bascom and his colleagues quit and returned to Washington. A Brown & Root official said, "They had nothing to teach us." Bascom's view was, "We had everything to teach them. They just didn't want to listen." In a swift turnabout, Bascom's OSE then became consultants to NSF, an arrangement which wore through after ten months. At that point then, Bascom forever departed from any official connection with Project Mohole, after having served as director, then as consultant to the contractor, and finally, as consultant to the employer of the contractor. OSE struck out on its own. Started in 1962 with a capitalization of $9500, it was doing $3 million worth of oceanographic business (such as exploring for diamonds off South Africa) by 1966, and listed its net worth as $4.5 million.[22] Meanwhile, AMSOC, under Hedberg's prodding, had lined up behind the two-ship, two-program concept. And on Capitol Hill, Senator Gordon Allott, of Colorado, the senior Republican on the subcommittee that handles NSF's budget, was angrily attacking the selection of Brown & Root over Socony Mobil, whose proposal NSF had once considered "in a class by itself." The Senator's interest in the case provoked Brown & Root to an assortment of gyrations. It was repeatedly suggested, for example, that Allott's interest in Mohole originated with disappointed constituents who had sought the contract. Allott emphatically denied this, pointing out that, as far as he knew, his state's only connection with Mohole was through a Brown & Root subsidiary in Colorado.[23] As the Senator bore down on the project, Brown & Root sought channels of silencing him. On May 16, 1963, Brown Booth, director of public relations for Brown & Root, wrote to Robert Six, president of Continental Airlines, which is based in Denver, "At the suggestion of Joe Thomas,* one of our directors, I am writing you for advice and counsel about a problem we are having with our Project Mohole. He understands you are a friend of Senator Allott." Booth stated that Allott apparently was unaware of Brown & Root's distinguished record, and, as a result of his attacks on the contract award to the firm, "for the first time the honorable name and reputation we have built up over forty-seven years of hard work is taking an unmerciful public drubbing." "If we could reach the Senator, I believe we could convince him that Mohole is in good hands and that excellent progress is being made. . . . We think we are Senator Allott's 'kind of folks' and that if he knew us he would be with us."

Booth's letter to the president of the Colorado-based airline was forwarded to Allott by the airline's director of public relations. A few days later, Allott replied, "I think this is about the sixth or seventh communication I have had of a similar nature, and still the representatives of Brown & Root have made no attempt to get in touch with me personally." Allott continued to attack the management of the project; meanwhile, wholly independent of the Senator's efforts, the ingredients for misfortune were accumulating at a rapid pace.

* No relation of Representative Albert Thomas.

❖   ❖   ❖

Early in 1963, while Brown & Root was preparing the preliminary designs for fulfilling its contract with NSF, Hedberg assembled the AMSOC executive committee for the purpose of firmly establishing that body's policy on the scope and objective of the project. The effect, however, was a further compounding of confusion. Under Hedberg's leadership, the executive committee tied itself to the chairman's broad-scale concept of a two-ship, two-program approach. But then Hedberg, in forwarding the committee's recommendations to Frederick Seitz, the newly installed president of the Academy, added the following,

"It is certainly my own strong feeling that *this experimental-exploratory stage* (sometimes called 'intermediate stage') *must be carried out as an integral part of the AMSOC project* [original italics], since it appears highly desirable for the attainment of the ultimate objectives of the project and since in my opinion the achievements to be expected from this stage are necessary to the justification of the whole project. *I do not think, however, that as a part of the AMSOC project it necessarily has to be carried out under the same contractor as the Mohole itself since the contract signed by NSF with Brown and Root, Inc., refers only to 'a hole through the crust of the earth* [italics supplied].' "

Seitz forwarded Hedberg's communication to NSF Director Waterman with the comment: "From my acquaintance with the extensive discussions of the scope and execution of the Mohole project, I am convinced that the recommendations of the executive group of our committee are sound, and I am glad to transmit them to you herewith." [24]

Thus, in early 1963, this was the status of Project Mohole: Hedberg, as chairman of the group which had originated Mohole, regarded the project as a broad and unrestricted two-ship drilling program, a program which seemingly had the endorsement of the National Academy of Sciences. Brown & Root, on the other hand, was working under a contract which directed it to devise a means for drilling to the mantle—and no more. And NSF, as author of the contract, presumably shared the Brown & Root conception, although in theory, Hedberg's committee was NSF's scientific adviser on the project.

One of Waterman's most frequently offered arguments for selecting Brown & Root was that, though the Texas firm did not itself possess the technical staff for carrying out the project, it had the reputation, resources, and managerial ability to acquire and direct such a staff.

But as the months rolled on following the award of the contract, Brown & Root's performance did little to justify Waterman's confidence. The firm's first choice for manager of the Mohole project was suddenly switched to another plum that had fallen to Brown & Root—construction of NASA's Manned Spacecraft Center, in Houston. In May 1963, *Fortune* reported: "In little over a year Project Mohole has had four other titular or *de facto* managers. Though pay has been higher than at Brown & Root generally, recruiting has been slow and turnover has been high." [25] At one point, Senator Kuchel of California took to the floor to lambaste Brown & Root's performance. The Texas firm's public relations director replied, "Certainly our project manager, Bowman Thomas, has had more experience in drilling offshore than any other human being. I presume the Foundation considered this in its decision to give us the contract." Three months later, Project Manager Thomas resigned from Brown & Root. Toward the end of February, Senator Allott returned from a luncheon meeting with Jerome B. Wiesner, science adviser to President Kennedy, and dictated the following memorandum:

"Gist was that there is great doubt that Brown & Root will be able to fulfill contract—especially as regards trained personnel and scientific know-how. . . . We talked about many other things and he was close-mouthed—but indicated Waterman on way out * and that this matter was handled badly—that NSF just did not realize how big a thing this was and also that there was grave doubt if B & R have the scientific ability to [do] the job—a point I hammered on considerably."

Whatever the doubts about the ability of Brown & Root, in April, thirteen months after receiving the Mohole contract, the firm unveiled its recommendations for carrying out Phase Two of Project Mohole. The plan was spectacular—and so was Hedberg's reaction.

Brown & Root, under contract to devise a means for drilling *a* hole through the earth's crust, was under no obligation to tune into the one-ship two-ship controversy. As Waterman himself put it, in a statement that amounted to abdicating the Foundation's judgment to the Mohole contractor: "My view of the prime contractor's responsibility is that you give him the objective, and since he is expert at carrying

---

* Waterman, then 71 years old, and past the federal government's retirement age, was holding office under a special presidential dispensation while a search went on for his successor. Eventually the job went to Leland J. Haworth, an AEC commissioner who had formerly directed the Brookhaven National Laboratory.

out engineering operations, he takes into account all the advice you can give him, and then, since he is in charge of the operations, he should be the one to come up with the best plans as to how one does it." [26]

Brown & Root's scheme for "how one does it" was to go directly to the construction of the "ultimate" ship. And what a ship it was! The Mohole, Brown & Root proposed, would be drilled from a 10,000 ton floating platform, 234 by 250 feet, resting on six 29-foot-high columns, which, in turn, would rest on two submarine-shaped hulls, each 370 feet long and 35 feet in diameter. Driven by propellers on the stern of each submarine hull, the vessel would propel itself to the drilling site. Once there, the submarine hulls would be partially flooded, thus bringing the vessel's displacement up to 24,000 tons to provide greater stability in a pitching sea. For positioning the vessel over the drilling site, it would be surrounded by rings of sonar and radar buoys, both surface and subsurface. Signals from these buoys would automatically activate the propellers on the submarine hulls, plus propellers located in each of the supporting columns, and these would hold the platform on its drilling station, just as the outboard motors on the hull of the *CUSS I* had held that vessel. The positioning system, according to Brown & Root's calculations, would permit drilling to continue uninterrupted in even near gale-force winds of up to 33 knots.

Brown & Root estimated that the total cost of building the platform and drilling through the crust of the earth would be $67,700,000.

At this point, the Foundation's blundering performance was deemed intolerable by the fiscal and scientific advisers around President Kennedy. On March 1, 1963, Kermit Gordon, director of the Bureau of the Budget, forbade Waterman from making further financial commitments "until the situation is clarified." "You will recall," Gordon wrote, "that when this [the second] phase of the project was initially brought to our attention, total costs of $15 to $20 million were anticipated. Last fall, when a request for $15 million was included in the budget for further funding, a total cost in the neighborhood of $50 million was discussed. Since then, your congressional presentation for 1964 states that the Foundation regards $50 million as a minimum figure and that the ultimate total may be considerably higher." [27] At Wiesner's insistence, Waterman appointed a special five-man investigating committee, headed by a veteran troubleshooter in the ranks of the postwar statesmen of science, Emanuel R. Piore, for a brief period a member of

AMSOC, former chief scientist of ONR, and now a vice president of IBM. But meanwhile, Waterman, who had done an extraordinarily statesmanlike job in nursing the Foundation through a difficult infancy, accepted the budget figures with equanimity.

Depending on the point from which one started counting, the costs of Mohole had increased anywhere from three- to twelvefold. "Brown & Root has a very remarkable record in a number of ways," he explained when inquiring congressmen pointed to the burgeoning costs. "First of all, they are quite accustomed to contracts and know how to deal with them. Second, they do them on time and at cost. These are important considerations. They have always done a very skillful job." [28] Two years had passed since the successful completion of Phase One. Was the project proceeding at a satisfactory pace? a congressman asked. Again, the NSF director exuded confidence. "When one is in a position of wanting to make a wise and sound decision," he explained, "one wants to see to it that the evidence comes in promptly and that the decision is made when the evidence is in. This is the decision of the director of the Foundation in consultation with the Board. Under these circumstances, one doesn't like to fix an exact date because there is confidence in the decision-making machinery so that we will do it at the earliest possible time." [29]

Brown & Root was plainly living up to its end of the bargain. It had been hired by NSF to chart a plan for drilling through the crust of the earth—the contract explicitly stated that any other objective would be separately negotiated—and the firm had come up with a design for drilling through the crust of the earth. However, with plans now advancing to bypass AMSOC's proposal for an intermediate *ship,* and, presumably, also to bypass an intermediate drilling *program,* Hedberg deployed for battle. Having hammered away for an intermediate *ship* and *program* since succeeding Lill in 1962, Hedberg now presented the issue to the nineteen-member AMSOC committee in blunt terms. Leland Haworth, Waterman's newly appointed successor, he told the committee, had warned him that funds might be available for only one vessel—and no assurance could be given for later building a second vessel. Under these circumstances, Hedberg asked in a questionnaire to the committee, would AMSOC prefer "(a) to get the intermediate-size vessel built now and take its chances on getting the ultimate vessel later, or (b) to get the ultimate vessel built now and take its chances on getting the intermediate-size vessel later"? Twelve members voted for the intermediate vessel now; five for immediate construction of the ultimate vessel, and two did not return marked ballots.[30]

Thus, a majority of AMSOC was willing to gamble the project's future by starting with an intermediate program. Hedberg now drew attention to an Academy-Foundation agreement, concluded just a few months before, which stated that, while NSF retained final decision-making authority, "the project should be aimed to attain as far as possible the scientific objectives conceived for it by the AMSOC Committee of the National Academy of Sciences–National Research Council. It is understood that the NAS-NRC through its AMSOC Committee will provide these scientific objectives and this scientific guidance." [31]

Mohole had now turned into an enervating and seemingly interminable conflict for NSF. With ample supporting documentation, NSF could argue that it had come into Project Mohole with the understanding that it was financing a program to drill a hole to the mantle—and not a general program of multilevel drilling on land and sea. At least five of AMSOC's own members appeared to share this conception. And one month after Brown & Root had unveiled plans for the proposed platform, AMSOC's own drilling panel concluded: "It is our opinion that a properly designed floating drilling platform, with a suitable positioning system and adequate mobility, offers the best present solution of the requirements for both the intermediate and ultimate objectives of the Mohole project." [32] On the other hand, Captain Lewis A. Rupp (USN Ret.), a naval architect who chaired AMSOC's naval architecture panel, flatly stated, "I firmly believe that the public and the scientific community would be best served by carrying out a two-ship program. Immediate investment in a modest intermediate vessel, with deferral of construction of the ultimate vehicle until some of the development problems are better defined, would not only save the public considerable dollars, serve the scientific community more fully with earlier concrete results, but also minimize the risk of a major fiasco." [33]

Meanwhile, Bascom, now on the brink of financial success in general oceanography and undersea mining, was willing to make known his opinions on the progress of the Mohole venture. In response to an inquiry from the Bureau of the Budget, Bascom, on June 18, 1963, wrote: "Sixteen months after their selection Brown & Root shows little understanding of either the scientific or engineering problems. . . .

Recently we have reported to the Academy on their 'Engineering Plan' which contains none of the important elements usually contained in a plan. It is, in our opinion, totally unsatisfactory. . . . Nor is the NSF staff competent to manage the work or to make decisions on scientific engineering matters. They have not accepted our advice or that of the Academy committee and at this moment it appears that a huge expenditure of money is to be made to satisfy the whims of three or four NSF staffers. Brown & Root seems to be incapable of being efficient by our standards. They have already wasted over a year of time and millions of dollars and although they still have no plan they have asked that the budget be more than doubled. . . . NSF should release Brown & Root from the contract or, alternatively, if they must be kept because of their political influence, allow them to 'study' a large platform indefinitely at a rate of less than half a million dollars a year."

The science statesmen grappling with the botched affairs of Mohole were accustomed to, and could live with, such behind-the-scenes stiletto wielding, but what they could not endure was the rash of snickering at their ineptitude that began to appear in lay journals. In June *Newsweek,* for example, dealt with the debacle in an article titled "Project No Hole?" which asserted that "many top-ranking scientists have lost faith in Project Mohole." [34] The month before, *Fortune* had come out with "How NSF Got Lost in Mohole," a lengthy article that questioned the integrity and competence of the Foundation. For the leaders of science, with their traditional concern for maintaining an appearance of dignity and keeping disputes out of public view, Mohole had become an egregiously painful sore. What they did not realize was that it would become worse.

Policy making is essentially an ordering of priorities, an acceptance of the fact that, since resources are not infinite, it is not possible to pursue every objective that is physically within reach. An ordering of priorities, however, is repugnant to the ideology of science, and, in general, unacceptable to the statesmen of science. How can it be contended, they would argue, that knowledge of the chemical composition of Mars is more or less desirable than knowledge of the composition of the atomic nucleus? Furthermore, believing that basic research not only more than pays for itself in ultimately utilizable knowledge, but also is far from employed to capacity, why should they acknowledge a need for priority making? With this bit of background in mind, we can turn to the conclusions of the Piore panel, which was established by the Board of the Foundation to unravel the Mohole mess. In July 1963, with further funding of the project still prohibited by the Bureau of the Budget, the panel concluded that the way out of the impasse was simply to acknowledge the equality of both camps in the dispute over the intermediate versus the ultimate ship. "The panel feels strongly," it was reported in a memorandum to the National Science Board, "that the Mohole program should be prosecuted with great vigor, and that funds should be made available now for the construction of the neces- sary drilling vehicles. The panel unanimously urges that an intermediate drilling vehicle be constructed promptly, but that this should in no way impede the design and construction of the final vehicle by Brown & Root." [35]

Such was an easy, and predictable, way out for a part-time body of advisers. But for Leland J. Haworth, who, upon Waterman's retirement, had the misfortune to inherit the Mohole controversy, the political reali- ties surrounding the Foundation were such that he could not ask Con- gress to end the controversy by providing funds for both camps. How- ever, Haworth's solution, as had been the case throughout Mohole's contentious history, was to blur the issues. Appearing late in Novem- ber 1963 before a special session of the Senate Appropriations Com- mittee, he proposed that Brown & Root be permitted to proceed with the design and construction of the ultimate platform, but that the equipment on that platform be "something less than the ultimate drilling equipment." [36] The big platform, with the less-than-ultimate equipment, he said, would be used in an intermediate drilling program that would help develop equipment and techniques for drilling to the mantle. In- dependent of this, he went on, he favored what he called a "supple- mentary drilling program" that would not be a part of Project Mohole, though it might be carried out under the direction of AMSOC.

However, the break between NSF and the AMSOC leadership was now complete. Appearing before the same congressional committee, Hedberg proclaimed that "personally I would far rather see this project killed where it now stands than to see it carried out in a manner not worthy of its potentialities. . . . I have been told by some . . . that Congress and the public have been sold on the project as 'a hole through the Moho and that only,' so that is what it must be. This appears to me to be a most disappointing explanation, with its implication that the Congress is unable to think for itself." Hedberg, in a fiery mood, went on to assert that whatever the initial scheme might have been, it should now be recognized that the project should proceed in a step-by-step fashion, collecting information and experience along the way, before attempting to go on to the ultimate objective. ". . . It is not at all inconceivable," he warned, "that early results may indicate that there is either no need or no possibility of drilling to supposed Mohole depths, in which case it would have been a reckless disregard of taxpayers' money to have prematurely or needlessly built the huge vessel now proposed for Mohole drilling." The cost of using the ultimate platform for intermediate drilling, he estimated, "would alone amount to enough to go a long way toward paying the cost of the intermediate vessel." Hedberg concluded with a blast: "The initial false glamour of the Mohole idea is wearing off in the face of realities, and I am sure that the informed public now finds a much greater appeal in a broad sensible program of crustal investigation carried on at a moderate rate rather than in a crash Mohole stunt." [37]

In conventional politics, Hedberg's forceful rhetoric would probably have elicited no great notice, but the AMSOC chairman was not engaged in conventional politics; rather he was engaged in science politics, in which a primary rule of conduct is decorum before the nonscientific world.

Two days after Hedberg's congressional appearance, he was in receipt of a letter from Academy President Seitz, which opened with a deceptively cordial "Dear Hollis," and then went on to convey a frosty rebuke. "It appears that at some time in the last month," Seitz wrote, "you sent Dr. Haworth a letter on Academy stationery spelling out the Academy's policy on the Mohole Project as you see it, without first clearing the letter with me. Had you discussed the letter with me, I probably would have asked you not to send it without having it reviewed by the [Academy's] Council because I understand it contained admonitions concerning what you would say in public if your point of view was not followed in detail by Dr. Haworth in his future negotiations regarding the Mohole program." Seitz then proceeded to chastise Hedberg for his congressional testimony, accusing him of presenting "your personal views concerning the course of the Mohole Project . . . as if they might be the official views of the Academy. . . . Whatever the merits of your point of view may be," Seitz continued, "I feel it was improper to present such formal testimony to the Congress without clearing your proposed testimony with me so that I might have referred it to the Council of the Academy if I considered such a step wise." The testimony, Seitz wrote, did "not give adequate recognition to the nearly superhuman effort the new director of the National Science Foundation has been making to find a workable compromise among the various controversial points of view regarding the Mohole." And then, the ultimate revelation of Seitz's chagrin: "Your testimony serves to make his already difficult task much harder and hence, in my opinion, reflects badly on the Academy as an advisory organization. . . . AMSOC has benefited in stature from the fact that it is part of the Academy and must expect to live within a framework of conduct designed to protect the reputation of the Academy."

Accompanying this admonition was a delicately phrased but unmistakable ultimatum: "I would like to inquire if you, as the current chairman of AMSOC, are prepared to assure me that henceforth you will not communicate with any organization or agency outside the Academy . . . on matters of major policy regarding the Mohole Project without first consulting me. . . . Unless you can assure me on this point, I will have no choice but to request the Council to permit me to reconsider your own status as chairman of AMSOC."

Seitz's letter elicited an angry response from Hedberg—on Academy stationery. Submitting his resignation as AMSOC chairman, effective immediately, Hedberg told the Academy president that "I cannot help but find your letter not only inaccurate but most unfair and misleading in the impression it gives." He described as "patently ridiculous" Seitz's assertion that his letter to Haworth should be interpreted as "spelling out the Academy's policy on the Mohole project." The letter to Haworth was necessary, he said, because Haworth was unable to see him prior to his congressional appearance. Hedberg said he had marked the letter "Private and Confidential" and had not distributed any copies, even to the AMSOC staff. He added, however, "I must assume that you saw the letter, even though you have recently asked me for a copy," to which he added that he had advised Seitz's office of his intention to respond to the congressional invitation to testify, but that Seitz was preoccupied with preparations for the Academy's centennial observation. Hedberg concluded, "It seems to me that the stature of the National Academy and its reputation is vitally tied to the freedom with which its members may express themselves on matters of scientific import. . . . I also trust that some of the hysteria which seems to be surrounding this Mohole Project will soon be dispelled under wise leadership by you and Dr. Haworth."

Two years before, Mohole had lost the services of the Bascom group, which had brilliantly carried out the first phase of the project; now it lost the services of Hollis D. Hedberg, who, despite his acerbity, had done more than anyone else in that strange cast of characters to clarify the muddled thinking that had characterized Mohole from its day of conception. To proceed with the venture, it had become necessary to jettison some of those very few people who had the ability and foresight to make Mohole work.

By early 1964, Seitz and Haworth had rooted out the dissenters, and all was seemingly well with Mohole. Seitz would no longer have AMSOC on the Academy's prestigious premises, and it therewith dissolved, with its place in the project being taken by two newly established Academy committees, one for dealing with the project's scientific aspects, the other with engineering matters. At the Foundation, Gordon Lill, co-originator of AMSOC, reappeared from his stay with Lockheed to take up the newly established post of Mohole project director. On January 21, Haworth held a press conference at which he announced that the Bureau of the Budget had lifted the ban on further expenditures. Despite the vast contention and the mounting estimates, relatively little had so far been actually expended. Phase One had cost a total of $1.8 million. And by the end of June 1963, only $4.9 million had been spent or contracted for on Phase Two. But now, Haworth said, progress and expenditures would rapidly accelerate. Fiscal 1964 would require expenditures and obligations of nearly $8 million, and the figure for the following year would be approximately $25 million. The total cost—design, construction, and operations—for reaching the mantle would be approximately $68 million, he said, but Haworth advised his audience to drop the notion of Mohole as a one-shot affair. ". . . You just have to understand and everybody has to understand that it just isn't the way you do it. You don't include in the cost of an aircraft carrier the cost of the first battle but no other battle. . . . You are not going to build a facility that costs $30 million or $40 million or whatever million dollars and do one job with it and dry-dock it." [38]

The appearance now was one of stability, but the reality was that Mohole was running without the ingredients necessary for success in any major research venture: broad political support (or at least the absence of significant opposition), and the interest, enthusiasm, and dedication of scientists who command the respect of their colleagues. In the House, Representative Albert Thomas stood alone as an influential champion of the project; there was no similar supporter in the Senate. Meanwhile in both houses, while the rank and file were indifferent to the project, many members held grudges because of the questionable circumstances that surrounded the contract award. In the scientific community, the originators of the project continued to regard it as important, but not so important as to warrant their devoting significant time to it. Munk and Hess had little to do with it after the first few years; when the Foundation sought to place the project in the hands of a consortium of universities—which was its pattern of operations with other large facilities, such as its radio astronomy center in West Virginia—there were no takers. While the beast sired by NSF was gestating in Brown & Root, the universities preferred to remain uncommitted.

Despite the aloofness of academic science, a period of equanimity soon came to prevail in Project Mohole. Early in 1965, NSF announced the tentative selection of a drilling site near Hawaii. By the fall of that year, Brown & Root's belatedly completed designs passed scrutiny, and NSF gave the Texas firm authority to invite bids for construction of the platform. Two years earlier, Haworth told Congress that Brown & Root's construction-cost estimates ranged from $18 million to $25 million.[39] The National Steel and Shipbuilding Company of San Diego was awarded the contract with the lowest of four bids—$26.9 million; the highest was a trifle over $45 million.* With the construction of a seagoing platform that could drill through the crust of the earth, Phase Two of Project Mohole was at last taking material form.

* Waterman, in one of his last congressional appearances before retiring, presented Thomas with a report that said the Foundation had not yet decided whether to build a new ship or convert an existing one for drilling to the mantle. The report stated, "The present estimate is that the ship will cost from $11 million to $12 million. . . ." Brown & Root, it added, estimated the total cost—construction and drilling—at $50 million. But NSF's report to Thomas added the caution, "The Foundation believes that $50 million is a minimum figure and that the ultimate total may be considerably higher." Upon reading this, Thomas, the veteran budget-cutter, said, "And doubtless will be. It would not be true to form if it was not." (*House Independent Offices Appropriations, Hearings,* 1964, p. 489.) The fact of the matter was that the costs were beyond accurate estimate in the days before the project was clearly defined. But, since Congress demanded firm figures, the Mohole proponents supplied them.

In January 1966 Representative Albert Thomas, the omnipotent chairman of the House Independent Offices Subcommittee, died of cancer. The first major action of his successor, Representative Joe L. Evins, Democrat of Tennessee, was to deny further funds for Project Mohole. "In view of the current world situation," Evins' subcommittee stated, "and the need to continually review priorities, the Committee has not allowed funds for Project Mohole. This project has progressed slowly with considerable difficulty—the estimated cost is in excess of $75,000,000. The cost of the project has greatly exceeded the original estimate and promises to increase still further. The Committee suggests that the funds of the Foundation can be more advantageously used in other activities and no funds are included to continue this project."

Haworth and a hastily collected rescue party, including Hawaii's two senators, appealed to the Senate to salvage the project. The Senate made a feeble attempt to reverse the House decision, but Mohole had no friends in the Capitol, and it had many enemies. In hearings in the Senate, NSF officials now spoke of costs of at least $125 million, and pleaded that the project be permitted to proceed. The Senate was willing to go along, but the House stood adamant. For a time, the fate of Mohole hung in the balance, but then, quite suddenly, and for reasons altogether irrelevant to the merits of the project, the balance swung decisively against Mohole. On August 18, Representative Donald Rumsfeld, Republican of Illinois, revealed that Brown & Root's chairman George Brown, and members of his family, had contributed a total of $23,000 to the President's Club, a Democratic fund-raising organization. The contributions came a few days after President Johnson had successfully appealed to the Senate to reverse the House decision to kill the Mohole project. Since persons associated with Brown & Root had long been suppliers of funds for Lyndon Johnson and the Democratic Party, and Brown & Root at the time held several hundred million dollars' worth of federal construction contracts in Vietnam and elsewhere, it is doubtful that the timing of the donations was any more than a coincidence, or that they bore any relationship to Johnson's appeal in behalf of Mohole. But, in Congress, the political base of the project was now virtually shattered. And the Rumsfeld revelations, irrelevant as they were to the question of whether Mohole should be permitted to proceed, added a weight that brought the entire venture crashing down.

On August 24, 1966, it was noted in the Congressional Record that the "Senate adopted conference report of H.R. 14921 . . . receding from its disagreement with House amendment No. 26 (elimination of funds for Mohole project). These actions cleared bill for President's signature." On the following day, not quite a decade after Walter Munk suggested that a hole be drilled through the crust of the earth, the National Science Foundation issued an announcement titled, "NSF Initiates Action to Stop Project Mohole." Mohole was dead.

NOTES

1. Harry H. Hess, "The Mohole Project, Phase II," *Drilling for Scientific Purposes*, Report of Symposium, Geological Survey of Canada, Paper 66-13 (Ottawa: Queen's Printer and Controller of Stationery, 1966).
2. Willard Bascom, *A Hole in the Bottom of the Sea* (New York: Doubleday & Co., Inc., 1961), p. 47.
3. Quoted in Willard Bascom, "The Mohole," *Scientific American*, April 1959.
4. Bascom, *A Hole in the Bottom of the Sea*, p. 50.
5. Reprinted in *Mohole Project, Hearings*, Subcommittee on Oceanography, House Committee on Merchant Marine and Fisheries, 1963, p. 162.
6. Gordon G. Lill and Arthur E. Maxwell, "The Earth's Mantle," *Science*, May 22, 1959.
7. Bascom, "The Mohole."
8. Summary of AMSOC Executive Committee minutes, "Mohole Project Chronology," mimeographed, National Academy of Sciences.
9. Quoted in *Science*, January 10, 1964, p. 118.
10. NAS-NRC Publication 914, 1961.
11. *Mohole Project, Hearings*, p. 171.
12. *Ibid.*, pp. 171–72.
13. Summary of Drilling Techniques Panel minutes, "Mohole Project Chronology."
14. *Mohole Project, Hearings*, p. 209.
15. *Ibid.*, p. 259.
16. Herbert Solow, "How NSF Got Lost in Mohole," *Fortune*, May 1963.
17. Quoted in Report of the Comptroller General of the United States, June 18, 1962, *Senate Independent Offices Appropriations, Hearings*, 1967, pp. 1777–81.
18. Solow, *op. cit.*
19. Report of the Comptroller General.
20. *House Independent Offices Appropriations, Hearings*, 1963, p. 855.
21. *Mohole Project, Hearings*, pp. 182–86.
22. Paul O'Neill, "Trailbreaker of the Deeps," *Life*, September 30, 1966.
23. *Congressional Record*, August 10, 1966, p. 18084.
24. *Mohole Project, Hearings*, pp. 190–91.
25. Solow, *op. cit.*
26. *Mohole Project, Hearings*, p. 14.
27. *Senate Independent Offices Appropriations, Hearings*, 1964, p. 2348.
28. *Mohole Project, Hearings*, p. 33.

29. *Ibid.,* p. 18.
30. *Ibid.,* p. 85.
31. Agreement Between the National Science Foundation and the National Academy of Sciences–National Research Council on the Role of the AMSOC Committee and the Scientific Objectives of Phase II of Project Mohole, March 18, 1963.
32. *Mohole Project, Hearings,* p. 220.
33. *Ibid.,* p. 77.
34. *Newsweek,* June 10, 1963, pp. 97–99.
35. *Senate Independent Offices Appropriations, Hearings,* 1964, p. 2376.
36. *Ibid.,* p. 2362.
37. *Ibid.,* pp. 1638–47.
38. Press conference, National Science Foundation, January 21, 1964, mimeographed transcript.
39. *Senate Independent Offices Appropriations, Hearings,* 1964, p. 2336.

X
High Energy Politics

In my opinion, the necessity for large and expensive accelerators is a very sad necessity which nature seems to be forcing upon scientists.

NORMAN F. RAMSEY, Harvard Professor of
Physics, Testimony to Joint Committee
on Atomic Energy, 1959

I told them that it was a great mistake to take the political route, and they said, "That's all we can do, we don't have any Nobel prize winners on our side."

A West Coast Physicist's Report
of a Conversation with a Midwestern
Colleague, 1966

The financial difficulties of the early 1960's inspired the chemists to collaborate in stating their case for federal support. As practitioners of a nationally dispersed "little science," the chemists prudently concluded that unity was the fruitful path to relief. However, their aristocratic "big science" colleagues, the particle physicists, responded to the drought differently. Seeking out alliances with congressmen and governors, they fell to battling each other with vigor and acerbity of such magnitude that by late 1963 Lyndon Johnson, less than one month in the White House, felt compelled to intervene. When the internecine strife had at last been adjudicated, Johnson stated to Hubert Humphrey, a senatorial champion of one of the physicist factions, "I devoted more personal time to this problem than to any nondefense question that came up during the budget process." [1] * Allowing for political hyper-

---

* Amidst the grief and uncertainty that followed the assassination, this episode, though unprecedented in pre- or postwar dealings between science and government, attracted relatively little public notice. In the most placid of times, the politics of science is generally ignored by the lay press, but in the agitated atmosphere of the immediate post-assassination period, an issue arising from a seemingly opaque squabble among scientists was not likely to receive significant notice.

bole, there is no doubt that the newly installed President spent a great deal of time plumbing the political intricacies of particle physics. Nor is there any doubt that Johnson's intervention marked the beginning of a major transformation in the politics of science. Though basic science had never been given wholly free rein by its federal subsidizers, the politicians' attitude had theretofore ranged between enthusiasm, or permissiveness, at best, and indifference at worst. In the Age of Johnson, however, science was deemed both too important and too expensive to be left exclusively to scientists. And, from that point on there developed a new phase in the politics of science, one in which the power of the long-reigning statesmen of science was substantially reduced, and nonscientific values increasingly intruded into the preserves that had once been governed almost exclusively by men of science.

To examine the genesis, development, and effects of this change, let us now look at the field of research that provided the precipitating event for Johnson's intervention: high energy physics—at once the costliest, least intelligible, and least utilitarian in the entire spectrum of basic research. Virtually untouched by federal subsidy prior to World War II, high energy physics had burgeoned to a position of $143 million a year in the federal budget by 1963, and it was in that same year that high energy physicists were laying plans that, if accepted, would cost the federal government $600 million a year by 1981.[2] It was those plans that touched off the fight.

In 1895 Roentgen discovered X rays, and the following year, Becquerel discovered radioactivity in a sample of pitchblende. The atomic revolution was born of these two events. The Curies subsequently obtained a fairly intense source of radioactivity by separating radium from pitchblende. Studies of radioactivity then led to the discovery of three distinct types of radiation, which were designated alpha, beta, and gamma rays. Using these findings, scientists proceeded to probe still further. And, in the best traditions of basic research, each step brought into view new mysteries which, in turn, usually necessitated new equipment for new explorations. In 1909, Ernest Rutherford and other scientists in Britain sent an electrical charge through alpha particles which had previously been collected in a vacuum chamber. From this experiment they concluded that the particles were actually helium atoms minus two electrons. Between 1911 and 1913, Rutherford and his associates demonstrated that the mass of the atom was concentrated in an in-finitely minuscule region called the nucleus.* Niels Bohr then conceived a planetary model of the atom in which the nucleus is surrounded by orbiting electrons. Years of effort were then directed to probing deeper into the atom and its components. Since natural radioactivity provided only limited energies, physicists in Britain and the United States, dur-ing the 1920's and 1930's, began to develop machines, known as ac-celerators, or "atom smashers," to impart to atomic particles the en-ergies required for effectively bombarding the atom.

In 1930, two British physicists, John D. Cockcroft and Ernest T. S. Walton, achieved the first nuclear disintegration produced by artificially accelerating protons within the nucleus. The potential of the Cockcroft-Walton accelerator was limited, however, by its large voltage require-ments. In that same year, Ernest O. Lawrence, working at the Univer-sity of California at Berkeley, developed the magnetic resonance ac-celerator, which he called the cyclotron, for imparting very great veloc-ites to heavy nuclear particles without employing high voltages. For the politics of physics, the overwhelmingly important fact was that research on this frontier now necessitated huge and expensive apparatus. Now the physicists were advancing into a field in which massive sums would be required. Lawrence's first machine weighed eighty-five tons, and eventually yielded particles with an energy of eight million electron volts (mev.), the volt being a convenient measure of the energy im-parted to elementary particles. His achievements on this frontier of research brought him the Nobel Prize in 1939, but from the point of view of the development of the politics of science, Lawrence's achieve-ments brought something even more significant: some of the most bril-liant young scientific minds in the country quickly flocked around him to be initiated into the exciting new field of particle physics. And be-cause they were bright, and because they knew physics, they quickly rose high in the wartime research effort, and with the wartime experi-ence behind them, they naturally flowed into the high postwar councils of science and government. Among these students and associates were

---

* Rutherford's work was a landmark on the course that eventually led to nuclear fission. But he himself took a modest view of the implications of his findings. In 1933, he said, "Anyone who expects a source of power from the transformation of these atoms is talking moonshine." (New York *Herald-Tribune,* September 12, 1933.)

J. Robert Oppenheimer, Glenn Seaborg, Luis Alvarez, Philip Abelson, and Edwin M. McMillan.

When World War II began, Lawrence was embarked on what has remained to this day a never-ending quest of particle physicists: higher and higher energies necessitating, of course, bigger and ever costlier machines. Until World War II, his work had been funded without federal assistance, with the exception of some minor aid from the Works Progress Administration.*

At the end of World War II, General Groves, eager to establish a peacetime atomic research program, provided Lawrence with $170,000 to complete a still more powerful machine, the 184-inch 200-mev. synchrocyclotron. This was far and away the most powerful accelerator in existence, which, in turn, meant that Berkeley, at the western extreme of the country, held a guarantee of dominance in research on the most exciting frontier of physical science. Such a situation was, of course, intolerable for another Nobel laureate in physics, Columbia's strong-willed and ambitious I. I. Rabi, who, though not a high-energy physicist, was a high-energy politician of science, and one committed to a postwar renaissance of physics in the East. Both Lawrence and Rabi had occupied key positions in the wartime research effort, Lawrence as an influential advocate of an all-out attempt to build the atomic bomb and later as director of the electromagnetic separation process for producing U-235; and Rabi as number-two man at the MIT Radiation Laboratory. After the war, Lawrence, in faraway California, did not wholly absent himself from the Washington scene, but his principal occupation was the design and construction of ever more powerful accelerators. Rabi, on the other hand, though returning to Columbia,

---

* See Chapter III, p. 65. Amidst current controversies over whether federal research policies insure that the rich get undeservedly richer, it is worth noting the economic genesis of Berkeley's distinction in high energy physics: The University of California achieved distinction in this field with is own funds, long prior to the arrival of federal subsidies for university-conducted research. It is true that other universities, among them Wisconsin, Illinois, and Rochester, closely followed in California's tracks, also with nonfederal funds. But California was the pioneering center in high energy physics, and the regents of the University were willing to dip into funds controlled at their discretion to enable Lawrence to maintain his leadership. After the war, when federal money was forthcoming, it was sensible, if not inevitable, that federal research administrators would seek to build upon existing distinction, rather than try to make the deserts bloom. Thus, the rich did get richer, but it cannot reasonably be contended that they were undeserving of their happy fate.

became a central figure in the ranks of the northeastern scientists and science administrators who regularly commuted to Washington to conduct the affairs of science and government. Against this background, a cheerful competition for high energy eminence developed between Lawrence's Radiation Laboratory at Berkeley and the Rabi-inspired Brookhaven National Laboratory on Long Island.* However, since high energy physics was one of the most exciting frontiers of science, and physicists, including many of Lawrence's and Rabi's colleagues and former students, predominated in the councils of science and government, this competition between the eastern and western giants of physical research took place in a context of considerable affluence and general cooperation. The Office of Naval Research, with Alan T. Waterman, a physicist, as its chief scientist, and other physicists dominating its advisory councils, took the financial lead in the period between the decline of the Manhattan Project and the arrival of the Atomic Energy Commission. In fiscal 1947, for example, ONR, with a total research budget of slightly over $13 million, allocated $4.5 million to high energy physics. Later, the AEC came in with even greater support and by 1950 federal support in this field had risen to $12.5 million.[3]

The only blemish in this happy picture of coastal competition was the more or less loosely defined region known as the Midwest.† Prior to World War II, an ample store of resentment had accumulated between the University of Chicago, which was the prestigious private institution of the region, and the many state universities in the surrounding area. Disdaining the football-playing "cow colleges" of the region, offering superior salaries, and regarding itself as the unique embodiment of the eastern university tradition in the barren Midwest, Chicago radiated toward its neighbors a feeling that they were academic second-raters. And they, naturally, responded with a sense of resent-

---

* Lawrence called his nuclear physics research center the Radiation Laboratory; upon his death, it was renamed the Lawrence Radiation Laboratory. The electronics and radar laboratory established at MIT during World War II was also called the Radiation Laboratory; after the war it was rechristened the Research Laboratory for Electronics.

† One definition of the term is to be found in the membership of the Midwestern Governors' Conference, in which are represented Illinois, Indiana, Iowa, Kansas, Michigan, Minnesota, Missouri, Nebraska, North Dakota, Ohio, South Dakota, and Wisconsin.

ment toward the presumptuous and supercilious private institution that denigrated their intellectual legitimacy.

After the war, the Argonne laboratory, which had been developed at Chicago through Compton's machinations,* was designated a national laboratory of the Atomic Energy Commission. In theory, qualified researchers from universities throughout the country were to have equal access to Argonne's federally paid for facilities. But, for purposes of management, Argonne was put under the administration of Chicago, an arrangement which immediately heightened the old-time tensions between Chicago and other midwestern institutions. The animosities were in no way lessened when it was learned that the AEC would be paying Chicago a "management fee," above and beyond the direct costs of operating the Argonne facility. Eventually this fee reached the neighborhood of $2 million a year, a sum equivalent to the return on $50 million in endowment invested at 4 percent. On top of this, in sorting out various responsibilities the AEC decided that Argonne, which during the war had pioneered in the development of reactors for the production of plutonium, should continue this work into peacetime for the purpose of developing reactors for the production of electric power.† Accordingly, the AEC appointed to the Argonne directorship Walter H. Zinn, a physicist whose career had been devoted to research relevant to designing and building reactors for releasing the energy of the atom. Many midwestern physicists were just as interested in high energy physics as were their colleagues on the East and West coasts. But there was Argonne, the only major AEC facility in the region, assigned by the AEC to concentrate on the useful but scientifically mundane task of producing electricity from the atom. These were the waning days of the Manhattan Project, and General Groves was doling out vast sums that were to influence regional patterns of academic research for decades to come. Argonne, with a mandate to concentrate on reactor research, was allotted $5 million; the proposed Brookhaven Laboratory, with no such restriction on its efforts, was given $9.4 million; the Radiation Laboratory at Berkeley was allotted $2.5 million, and several million dollars were divided among the University of Washington, the

---

* Chapter VI, p. 100.

† To the lay observer, reactor physics might not appear very different from high energy physics. But in terms of equipment, objectives, and personnel, they are altogether distinct activities, sharing little more than common roots in physical theory.

University of Rochester, Iowa State, Columbia, MIT, and the Battelle Memorial Institute.[4] Thus, while the Midwest received a generous share of these funds, its share was largely restricted to reactor research. Meanwhile, amidst the tensions that traditionally existed between the University of Chicago and its neighbors, the following national pattern began to emerge in the glamorous field of high energy physics:

In 1948, Brookhaven, in the East, started construction of the Cosmotron, a 3-bev. proton accelerator, that went into operation in 1952. The construction cost was $9.3 million. In 1949, Lawrence, in California, began construction of the Bevatron, a gigantic, factory-size machine with an energy of 6.2 bev. This was completed in 1953 at a cost of $9.7 million. In addition, a number of smaller machines were spread among various universities; however, no major machine was allotted to the Midwest.*

By 1953, with Argonne assigned to concentrate on reactor research, fifteen midwestern universities, including the University of Chicago— whose physicists had their own differences with the Argonne bureaucracy—established a consortium, the Midwestern Universities Research Association (MURA), for the sole purpose of building and operating a large particle accelerator in the Midwest.† Financed by pledges of up to $150,000 from each member university (though the actual contributions were $10,000 each), MURA set up headquarters in a garage, near the Madison campus of the University of Wisconsin—and pro-

---

* In high energy physics, it is important to note, the initial construction costs, as large as they are, actually comprise only a minor portion of the total costs of particle research. An accelerator simply accelerates particles. To detect and study them huge and costly apparatus, such as bubble chambers and spark chambers, are employed in various configurations, depending on the nature of the experiment. As a consequence, a high energy physics laboratory resembles a factory that is forever being torn down and put together again. A rule of thumb for the economics of high energy physics is that annual operating costs generally run at least 20 percent of construction costs, and often a good deal more. Thus, the total construction cost is usually duplicated within the first few years of operation. As a consequence of the high operating costs, each existing machine swells the annual budget for the overall field and creates pressures against building new machines. At the same time, the users of the existing machines surround themselves with graduate students, many of whom go on to become particle researchers, thus swelling the ranks of would-be users.

† Ohio State, University of Michigan, Michigan State, Indiana University, Purdue University, Notre Dame, University of Illinois, Northwestern University, University of Chicago, University of Wisconsin, University of Minnesota, Washington University of St. Louis, State University of Iowa, Iowa State University, and University of Kansas.

ceeded to explore designs for a new type of accelerator, eventually to be known as the Fixed Field Alternating Gradient Synchrotron (FFAG). Though its energy would be relatively low—perhaps half of a giant new machine of at least 25 bev., the Alternating Gradient Synchrotron (AGS), then in the very early planning stages at Brookhaven—its designers hoped it would be uniquely high in the intensity, or the number per second, of particles it would produce. Thus, for the midwestern physicists who were dismayed by Argonne's lack of a major accelerator, MURA represented the hope for breaking into this scientifically exciting, costly, and prestigious field. One year after MURA's founding, this hope received sustenance from Washington in the form of two grants from the fledgling National Science Foundation, totaling $152,500. The following year, the Atomic Energy Commission came into the picture, providing MURA with $46,228 for design work, but more important, providing it with the expectation that once the uncertainties of the complex FFAG design were resolved, the AEC, far wealthier than the struggling NSF, would come forth with the huge sum required for building the MURA machine. In the early days, there were no clear-cut figures; the final figure arrived at, years later, was $170 million.

In 1955, while the MURA staff (which eventually reached 110 persons) was at work on designs to bring large-scale particle physics to the Midwest, there occurred the first major thaw in the Cold War. With the Korean War at an end, and both Eisenhower and Khrushchev inclining toward some sort of *modus vivendi,* the tradition of the internationality of science provided a convenient bridge for promoting peaceful East-West contacts. Out of the thaw came the Geneva Conference on the Peaceful Uses of Atomic Energy at which American scientists had virtually their first opportunity to hear their Soviet colleagues describe what they were up to in the new world of atomic research. What they heard, at least in the realm of particle physics, filled them with gross agitation, and thereby set in motion a complex series of events that ultimately led to the scuttling of MURA and the scientific-political blowup that was to reach the desk of Lyndon Johnson. It was the contention of the Russians that, at a high energy research center at Dubna, not far from Moscow, the finishing touches were being put on what would be the most powerful accelerator in existence, a 10-bev. synchrotron. When this came into operation, in a year or two, Russia would hold the lead in accelerator energy, at least until 1960 or 1961, when the machine being designed at Brookhaven was scheduled for completion.* In May 1956, these fears were substantiated during the first visits of American scientists to the Dubna center. Upon their return, they gushed with anxiety over what they interpreted, or at least said they interpreted, to be breakneck Soviet progress in the prestigious field of high energy physics. As one of them reported, the machine at Dubna, with a diameter of 700 feet, was an "awe-inspiring sight. It is almost twice as big as the Berkeley machine . . . costs and budgets don't count. Professor Veksler, who leads this construction, was asked frequently about the total cost of the machine. He shrugged his shoulders and said he didn't know. Veksler said, 'It was decided to build the machine, and then everything that is needed for it will be delivered.' " [5]

Los Alamos alumnus Robert E. Marshak, professor of physics at the University of Rochester, came back from Dubna to report: "It seems that scientific research in the Soviet Union is being pursued with an urgency which is reminiscent of a wartime operation. As my colleagues and I toured the nuclear research laboratories in the U.S.S.R. last May, we noted the same personal dedication to the task at hand, the same emphasis on speed rather than cost, the same unlimited financial support for facilities and equipment which we ourselves had known at Los Alamos during World War II.

"It seems clear that the scientific research program in the Soviet Union is getting enormous momentum and that the objective is to overtake American science in its great diversity, its high quality and its magnificent sweep." [6]

Ten years later, Marshak candidly conceded, "I am sure that the Russians would concur when I say that those optimistic views of the Russian potential in this field have not been fulfilled." [7] But in 1956, the appearance of oncoming Soviet hegemony on the most prized frontier of physics was not to be ignored.†

* Leland J. Haworth, who was then director of Brookhaven, says that word of Dubna had come west before 1955, but that no one took it very seriously. In view of this, the events that followed the Dubna revelation are all the more curious.

† Waving the red flag to stimulate the appropriation of funds for research and education comprises one of the less admirable parts of the postwar relationship between science and government. The practice can be explained, if not justified, on the grounds that, by and large, Congress and the Executive were more inclined to respond than to innovate, and that a Red scare was the only available device

One year prior to Geneva, the National Science Foundation had convened a panel of the statesmen of physics "in response to a general feeling that a review would be desirable as to the status of the high-energy accelerator situation." * The panel produced such recommendations as, "Government agencies should be receptive to proposals by qualified groups for the construction of high energy machines." Without specifying costs, except to point out that annual operating expenses of an accelerator were usually the equivalent of the average annual costs during design and construction, the panel recommended "construction of two to four accelerators in the range from 2 to 6 bev. be initiated in the next few years." "Interested and responsible" groups were indispensable for these great design and construction undertakings, it said, and such existed. Finally, the panel recommended that "no fixed general policy be made with regard to the location of new accelerators at individual universities, national laboratories or other research establishments"—all of which represented a feeling of satisfaction with the progress of federal subsidy, as well as a general indifference to the seemingly petty Argonne-MURA squabble.[8]

However, in 1956, four months after the opening of Dubna to American eyes, virtually the same panel was convened again. This time it concluded: "Although at present the United States is leading in the number and diversity of its facilities and in the productivity of its research programs, the rate of Soviet progress has been such that this situation may not persist through the next decade. . . . The great scientific interest in and the prestige that attaches to this field of research make it important for the United States to maintain its present position of leadership." The post-Dubna panel then got down to financial specifics. Noting that current annual expenditures for high energy physics

---

with which science could get the government's attention. In 1958, for example, NSF Director Waterman told a congressional committee: "All our scientists who go to Russia come back saying that the government is providing them [Soviet scientists] with the very best in facilities." (*Hearings,* House Independent Offices Appropriations Subcommittee, 1959, p. 322.) Four years later, however, Philip Handler, of Duke University, a member of the NSF Board, told the same committee that relative to the United States, the Soviets were generally backward in basic research. "Their biology effort is trivial, frankly," Handler said. (*Highlights of Science,* House Appropriations Committee, 1962, pp. 83–84.)

* The panel members were Samuel Allison, Chicago; Hans Bethe, Cornell; Leland J. Haworth, Brookhaven; W. K. H. Panofsky, Stanford; I. I. Rabi, Columbia; M. G. White, Princeton; John Williams, Minnesota; Jerrold Zacharias, MIT; and Robert F. Bacher, Caltech.

were $40 million, it recommended that by 1962, the level should be $60 million to $90 million in terms of 1956 purchasing power. As for the Argonne-MURA conflict, the panel repeated its 1954 opposition to establishing a "fixed general policy" on locating new accelerators. But now it was post-Dubna and plans were in the making. Accordingly, the panel went on to observe, "It is regrettable that no multi-bev. accelerator in operation or under construction is readily accessible to the large numbers of competent physicists in the Midwest. It is important that this deficiency be remedied by appropriate facilities." [9] But the panel made clear its belief that MURA was not yet ready to go into the construction stage. "The FFAG principle," it stated, "is at present in process of theoretical and experimental investigation." The work should proceed, but if construction were to start at once on a new accelerator in the Midwest, it could not be MURA's.

Since particle physics is remote from any application, military or otherwise, and basic science—as scientists regularly point out—is mankind's most cooperative endeavor, it is difficult to see why the prospect of a relatively brief Soviet energy lead should have inspired concern.* But in the best spirit of the Cold War, American physicists were now determined to embark on a crash construction program aimed at minimizing the duration of the Soviet lead. At that point, Berkeley was in the midst of exciting experiments with its new 6.2-bev. machine; Brookhaven had the Cosmotron to work with and was proceeding with the

* Even the most devout supporters of high energy physics readily acknowledge that the field is remote from any practical applications. Thus, Hans A. Bethe, in comparing the results of particle physics to the applications that have emerged from nuclear physics, candidly observes, "No such practical application has appeared, or is likely to appear, for particle physics." Bethe and many of his colleagues argue, however, that particle physics merits support because, as he puts it, it is "the most basic field of knowledge in the physical world," and its progress furnishes intellectual enrichment for other fields of science. There is no doubt, too, that particle physics is an excellent training ground for scientists and engineers who go off into other fields; furthermore, its technological requirements have been a great stimulus for developments in electronics and computer technology. (See the collection of essays in *Nature of Matter, Purposes of High Energy Physics,* edited by Luke C. L. Yuan, Brookhaven National Laboratory, 1965.) One of the more charming justifications for expenditures on high energy physics is offered in this collection by a scientist who ingenuously observes that this field of research "because of its financial magnitude and its eventual broad distribution of expended funds, can play a modest but not entirely negligible role in maintaining economic stability."

construction of the gigantic AGS. In the environs of the unhappy mid-western physicists, MURA was the only organizational aspirant for a major machine, but MURA was nowhere near ready to proceed with construction. As for Argonne, it was not particularly interested in high energy physics. Thus, a dilemma confronted the AEC, but, as it turned out, not an especially difficult one. On February 17, 1956, the way out was announced in a press release, titled "AEC Authorizes Two Accelerator Projects." [10] MURA, it stated, had been given authorization to design what the AEC hoped would be "a machine that will be the finest and most powerful in the world at the time of its completion. . . ." At the same time, the AEC announced, Argonne has been authorized "to design a high-energy accelerator which would permit earliest possible completion. . . ." Some of the details and rationale for this two-part solution emerged shortly afterwards at hearings before the Congressional Joint Committee on Atomic Energy. With AEC Commissioner Willard Libby, a scientific cold warrior if ever there was one, arguing the case, it was explained that Argonne would get to work at once on a simplified low-powered version of the gigantic machine that was being built at Brookhaven. Appearing before the Joint Committee early in 1956 to support a $15 million request for the proposed Argonne machine, Libby, in response to a question by Representative W. Sterling Cole (R-N.Y.), laid out the facts:

". . . Well, frankly, Mr. Cole, to get down to brass tacks, the thing . . . which the Russians showed us, is exactly in the energy range in which we think this [the proposed Argonne accelerator] is going to operate. And when their machine starts, the Russians are going to take over the show for several years. . . . The best we can do, I think, is to cut it down to two or three years. . . . Whatever we do, and it depends on the breaks even then, I am afraid they have got the jump on us a little bit. I am afraid that is it. They were very quiet about that machine. The first I ever heard of it was at the Geneva Conference." [11]

In the face of this impending energy gap, the Joint Committee, with little discussion and no dissent, authorized the expenditure of $15 million to construct a scaled-down version of the Brookhaven machine at Argonne. This, of course, came at a time when the leading physicists of the Midwest were staking their hopes on MURA as the organizational device for bringing high energy physics to their region. Nevertheless, MURA was not ready to proceed, and the AEC and the Joint

Committee were determined to get on with a program to cut down the Soviet lead.*

One year after the $15 million was authorized, but before any construction had begun, K. E. Fields, general manager of the AEC, and T. H. Johnson, director of the AEC division of research, reported to the Joint Committee that they felt it would be desirable to scrap plans for the $15 million machine and build a different type of accelerator, to be known as the Zero Gradient Synchrotron (ZGS), at a cost of $27 million. As things had developed, Johnson explained, the Soviet machine at Dubna "has not yet come into operation." (What Johnson did not know at that point was that the Dubna machine—the inspiration for a crash program to bring high energy physics to the Midwest—was turning out to be so great a technological fiasco that Soviet physicists eventually nicknamed it the Stalin Memorial Accelerator. It was not until the early 1960's that the Russians got the kinks out of it.) Now that sufficient time had elapsed for panic and opportunism to yield to the rational pursuit of good physics, it had been concluded that if a machine were to be built at Argonne, it might as well be a well-designed, useful one. Johnson explained: "The change in scope which is involved in the change from $15 million to $27 million was to get a machine of unique capabilities. The $15 million machine was based upon a scaledown of the machine that we are now building at Brookhaven, and that would have been a machine of too low an intensity to be really a useful research facility. So we have changed and now plan to build what we call a weak-focusing machine—it is a machine similar to the Bevatron at Berkeley—which will give us a high-intensity proton beam of 12½ billion volts. The Soviet machine . . . is 10 billion

* While American scientists became quite skilled at obtaining funds by sounding alarms about Soviet progress, there is evidence that Soviet scientists were not at all laggard in exploiting the Cold War to obtain funds from their government. Thus, Representative Melvin Price relates the following conversation with a Soviet physicist at the Dubna laboratory: "The Dubna Laboratory asked our group when we were there two years ago how we got the money to build our accelerators. We told him the legislative process of getting money on our program. He said, 'That is not the way I understand.' He said, 'I understand you get it by saying the Russians have a 10-million electron volt synchrotron and we need a 20-billion electron synchrotron and that is how you get your money.' I said, 'There may be something to it.' I said, 'How do you get your money?' He said, 'The same way.'" Price related this tale to John Williams, director of the AEC research division, who commented, "That is certainly a very true story." (*Stanford Linear Electron Accelerator, Hearings,* Joint Committee on Atomic Energy, 1959, p. 36.)

volts." [12] The congressmen went along without a quibble over this sudden near-doubling of costs, but even if they had chosen to quibble, the substance of the subject at hand was so far beyond their lay understanding that intelligent questioning, let alone dissent, would have been impossible.

"What is intensity?" Representative Cole asked Johnson, the research director.

"The number of protons accelerated per minute," Johnson replied. "It would be one thousand times greater than the $15-million machine we contemplated. That amount of current is needed to get satisfactory experimental results."

With humility, Cole sought to pursue the matter: "Not that it makes any difference if I know the answer, but will you tell me again, this will mean a greater intensity of what?"

"Of a beam that comes out of the machine of a larger number of protons per second."

"It increases the intensity of protons?" Cole asked.

"That is correct," replied Johnson, "and these are the particles you use for the research."

In an admission of his helplessness, Cole concluded, "Now I know what you are talking about but I do not understand it." [13]

The Joint Committee subsequently approved the request for $27 million. When the Argonne machine went into operation—in 1963— the total construction cost was officially placed somewhere in the low $40 million, but, according to one estimate, the actual figure was $51.4 million.

While the ZGS, accidentally fathered by the Cold War, was growing at Argonne, the future of the nearby MURA group looked relatively bright. After all, the decision to build a new accelerator at Argonne had been accompanied by the assignment of MURA to develop an accelerator that would be the "finest and most powerful in the world." Along with this assignment the AEC began to provide approximately $1 million a year to support MURA's staff as well as the construction of small working models embodying the FFAG principle. On the surface, all was harmonious, but behind the scenes, great difficulties were in the making. Later in 1956, MURA delivered to the AEC preliminary designs for the accelerator. What was proposed were two tangentially joined machines, in which particles would be accelerated and then brought into collision, thus amplifying the energy of the impact. The AEC, however, was skeptical as to the technical feasibility of the design, and asked MURA to develop it further. Meanwhile, however, the Dubna-inspired pressures were building up to get moving at once on the Argonne accelerator and the AEC proposed that the MURA group move to Argonne. The idea was that MURA's specialists could lend a hand with the Argonne machine, and then, when the kinks had been worked out of the MURA design, build their own machine on Argonne's spacious grounds. The motivation for this proposal was economy as well as speed. The Eisenhower administration was pledged to reduce the cost of government, and it seemed sensible to make all possible use of the existing facilities at Argonne rather than establish an altogether new laboratory in Wisconsin. But at that point, the animosities between Argonne and MURA had become so intense that lucid discussion, let alone a *rapprochement,* was impossible. MURA President Horace R. Crane, professor of physics at the University of Michigan, explained, "We have searched very hard for some possible organizational scheme which would allow the machine [MURA's FFAG] to be built at Argonne under partial MURA management or at Madison under some partial management of the AEC. . . . At two different times we had a meeting at the University of Chicago at which there were members of the Argonne Laboratory, and MURA representatives. . . . We searched very earnestly and everyone at the meeting was very sincerely desirous of finding some possible scheme. We came away from both of these meetings without any solution which we thought would work." [14]

The crux of the problem, of course, was that the MURA group did not wish to put itself at the administrative mercy of its old enemy, Argonne, while Argonne did not wish to grant sovereignty to any group on its premises, least of all a group whose very existence reflected dissatisfaction with the Argonne research program. But far deeper than these administrative wranglings, which are commonplace in and out of the scientific community, was the intensity of ill will that existed between the two organizations. I. I. Rabi, the eastern elder statesman of physics, reveals that in the early stages of the MURA-Argonne animosity he had attempted to serve as a mediator. As chairman of the General Advisory Committee of the AEC, and a leading advocate of high energy physics research, Rabi proposed that the MURA group agree to build its machine at Argonne. "Give us a chance to get it for you," Rabi says he told the MURA group, "and then when the question comes to where the actual machine will be built, it might go elsewhere." But, Rabi explained, "they had a hate on for Argonne. I found them [at MURA] so paranoid, so emotionally involved . . . so I quit. It wasn't my problem."

Corroboration of Rabi's assessment is to be found in a letter that Arthur Roberts, a senior staff physicist at Argonne, wrote to an AEC official several years later. It stated, in part: "Ever since MURA was set up, it has been scrupulously divorced from Argonne influence. The relations between MURA and Argonne in the early days of MURA can be described by no other term than war. The war centered around both personalities and issues; chief among the issues was the right of access of midwestern physicists to the facilities at Argonne. I have no knowledge of the details of this bitter, almost religious struggle. . . ."

Thus, in 1956, as a consequence of the spirit of Geneva and the opening of Dubna, it appeared that the barren Midwest was enroute to obtaining not one but two huge and costly high energy accelerators. The AEC was demonstrating its faith in the MURA group with a commitment of $1 million a year; Argonne was rapidly putting together designs for what was eventually to be the ZGS. A suddenly arrived era of plenty appeared to be the answer to the disharmony that existed between the two scientific groups. What was not realized, however, was that once again faraway events, this time on the West Coast, were shaping up to dim the new-risen hopes of the physicists in Wisconsin. To examine these events, let us briefly look back to a series of events that began two decades prior to the formation of MURA.

In 1934, after a year and a half at MIT and Michigan, W. W. Hansen, a twenty-five-year-old physicist, returned to his alma mater, Stanford University, to undertake a program of research into X-rays and other atomic phenomena. Since Lawrence, at nearby Berkeley, was concentrating on the acceleration of protons and other heavy particles inside the nucleus, Hansen, as is frequently the case in research, sought after unoccupied territory, and, accordingly, turned his attention to the task of accelerating electrons. His investigations led him to develop a device that came to be known as the rhumbatron, an empty copper container which could be made to act as a radio-frequency resonant circuit, capable of developing extremely high voltages with the consumption of only a limited level of radio-frequency power. However, it

soon appeared that Hansen's rhumbatron had been surpassed by another device, the betatron, developed by Donald W. Kerst, at Illinois.
Against this background of Hansen's fundamental and seemingly side-
tracked efforts, there was to occur one of those interplays of need and
knowledge that demonstrate the frequent arbitrariness and artificiality
of attempts to compartmentalize research as basic, applied, and developmental.

During Hansen's graduate days at Stanford, he became acquainted
with a fellow student, Russell Varian, who later worked as a researcher
for one of the pioneers of television, the Farnsworth Television Company. In the meantime, Russell's brother Sigurd had become a pilot
for Pan American Airways. In conversation with his physicist brother,
Sigurd expressed concern about the inadequacy of navigational aids
and instrument landing systems. With political upheavals wracking
Europe and growing talk of a new world war, he was also concerned
about means for detecting hostile aircraft. Russell concluded that the
solution to these problems might lie in the employment of radio
detection techniques. In those waning days of do-it-yourself research,
1934, the Varian brothers established their own laboratory, near Stanford, to pursue Russell's theory. They realized, however, that they were
up against some prodigiously difficult scientific problems and that they
would probably benefit from the scientific atmosphere at Stanford,
where, they knew, Hansen was tinkering with new microwave devices.
Hansen was receptive to an approach from his classmate, and, in 1937,
arrangements were made for the Varian brothers to conduct their research in the university's physics department. The university did not
pay them a salary but did provide $100 for materials and supplies.
The task at hand was to develop a device that could generate radio
waves in the centimeter region—waves sufficiently powerful to be
detected after reflection from a distant object. After several months
work, Russell had devised twenty-two different designs, all of which
Hansen rejected. Russell then came upon a principle which appeared to
answer the problem of very short wavelength radiation—the "transit-
time" effect. Hansen, though doubtful at first, agreed that the scheme
might be feasible, and work, greatly aided by Hansen's earlier efforts
on the rhumbatron, began on designing a prototype device of an ultra-
high-frequency electron tube soon to be known as the klystron. In
1940, with the acceleration of research into radar, the klystron came

to the attention of government and industrial researchers, and its development was speeded by the needs of war.

As in so many other fields of research, this war-accelerated translation of knowledge into hardware was to have a powerful feedback effect upon the scientists' capacity for pursuing still further knowledge. After the war, Hansen, E. L. Ginzton, and other members of the Stanford physics department returned to the task of accelerating electrons. (The Varians, meanwhile, had set up their own company, Varian Associates, for manufacturing klystrons and other electronic components.) Employing at first the war-developed electronic tubes known as magnetrons, Ginzton and his colleagues constructed a twelve-foot linear accelerator, a straight line device, as distinguished from the circumferential cyclotrons at Berkeley and elsewhere. In 1947, the machine known as the Stanford Mark I produced an electron energy of 6 million electron volts when powered by a single one-megawatt magnetron at a wavelength of 10 centimeters. But now greater energies were desired, and, furthermore, ONR was in the picture, eager to provide generous support for the pioneers in this field. Recalling their earlier work with the klystron, the Stanford group now sought to employ this device for accelerating electrons. In 1949, Ginzton and his associates fabricated and tested a new high-power klystron that demonstrated the practicability of attaining power ranging in the tens of millions of megawatts. ONR's Emanuel R. Piore, later to succeed Waterman as chief scientist of the Navy agency, wisely staked his faith in the Stanford group when it sought support for construction of a billion-volt accelerator. The machine, which was begun in 1949, successfully went into operation in 1952, one year before the MURA group came into being to bring high energy physics to the Midwest.

Meanwhile, the High Energy Physics Laboratory at Stanford had come under the direction of Wolfgang Kurt Hermann Panofsky, brilliant in physics and equally brilliant in the arcane politics of that trade. Under Panofsky the Stanford laboratory now began to lay plans for a still greater machine, what came to be known as the Stanford Linear Electron Accelerator (SLAC), one that would cost somewhere in the neighborhood of $100 million. In April 1957, when Argonne was beginning work on the ZGS, Stanford formally submitted to the Atomic Energy Commission a document titled "Proposal for a Two-Mile Linear Electron Accelerator." In support of the SLAC proposal, the Stanford group submitted a summary of a "technical assessment" performed by an outside group of scientists.* The group frankly stated that it was "worried that undertaking this project would result in inadequate support for other high energy facilties." It also characterized SLAC as employing a "brute force" approach to accelerator design, meaning that the proposed two-mile machine merely was a stretched-out version of earlier designs, rather than a new concept of accelerator design. But the summary also said the outside consultants had concluded that they "would prefer [SLAC] over MURA machine if choice is necessary." [15]

With SLAC suddenly arrived in the picture as a new competitor for the federal high energy research budget, the sad fact for MURA was that its design group still had not worked out the technical problems of the FFAG principle. Earlier that year, appraisals of the MURA design were sought from a group of scientists experienced in various aspects of high energy physics research. Among 37 who replied to questions on the feasibility of constructing MURA's designs, only 16 felt that feasibility had been established; 17 felt that the FFAG principle was questionable, while 5 felt that, though serious problems remained, the MURA staff could overcome them.[16]

MURA had been tested in the peer system and had failed to receive passing marks. SLAC, on the other hand, had aroused no questions as to feasibility, and now, the entire high energy community, including even the downtrodden MURA group, rallied behind Panofsky, director of the Stanford High Energy Physics Laboratory, to campaign for the approximately $100 million required to build his machine.† Panofsky's cause was undoubtedly aided by this remarkable show of unanimity, but it is not improbable that even if his colleagues had been unanimously against him, the remarkable Panofsky would nevertheless have triumphed. For in W. K. H. Panofsky, known as "Pief" throughout the

---

* Stanford refused to identify them, apparently on the grounds that to inspire frankness, an assurance of anonymity had been given to the consultants. This procedure is routine when scientists are asked to appraise the worthiness of papers submitted for publication. Whether it is desirable when vast public funds are involved is another question. Scientists accept without reservation the proposition that free and open discussion is the lifeblood of science, but when it comes to relations between society and tax-supported science, they frequently see virtue in settling their affairs behind closed doors.

† Appearing before the Joint Committee in July 1959, Rollefson, the director of MURA, generously stated that ". . . the Stanford and MURA programs are not competitors in the scientific sense, but will rather complement one another and will give more complete coverage of the high energy field than would be possible with either machine alone." As a representative of MURA, he said, he was offering his "enthusiastic support" for the SLAC proposal.

high-energy community, was epitomized a newly emerging generation in the elite ranks of the postwar politicians of science. In a sense, it might be said that he was the West Coast's young version of that venerable eastern statesman of science, I. I. Rabi. In 1942, when the renowned Rabi was associate director of the MIT Radiation Laboratory, Panofsky, with a Caltech Ph.D. in physics at the remarkably young age of twenty-three, served at his alma mater as director of an OSRD project in antiaircraft fire control. Later, he was a consultant at Los Alamos, where he worked with Luis Alvarez, one of the most inventive characters in the entire scientific community, on the development of devices for monitoring the effects of atomic blasts.* Following the war, while Rabi and many of the other senior leaders of the wartime research effort moved in the upper councils of the slowly developing science and government apparatus in Washington, the youthful Panofsky was establishing his reputation in physics, first at Lawrence's laboratory, where he joined Alvarez in building an accelerator, and then at Stanford, where he was appointed to a full professorship in physics in 1951. Two years later, he became director of Stanford's High Energy Physics Laboratory. Meanwhile, as was the case with so many of the brilliant young men of physics, he had begun to be drawn deep into the Washington advisory councils that stood at the junction of advanced physics and advanced weaponry.

In 1950, at the request of Lawrence, Panofsky had taken part in one of the lesser-known episodes of big postwar military technology: the crash construction at the Livermore atomic weapons laboratory of what was euphemistically referred to as the Materials Testing Ac-

---

* These devices were to be involved in a curious manifestation of the internationality of science. When the atomic attack on Nagasaki was being readied, Alvarez and several of his colleagues were concerned that the Japanese government might not fully recognize that a weapon of unprecedented destructive power was now being brought to bear on its cities. Recalling that a Japanese physicist, Ryokichi Sagane, of Tokyo University, had studied with them in 1938 under Lawrence at Berkeley, they addressed a note to him, urging him to advise his government of the significance of atomic weaponry. "As scientists," they wrote, "we deplore the use to which a beautiful discovery has been put, but we can assure you that unless Japan surrenders at once, this rain of atomic bombs will increase many fold in fury." The note was scotch-taped to a monitoring device that was parachuted over Nagasaki to transmit data on the blast effects. It was picked up thirty miles north of Nagasaki by the Japanese Naval Air Force, and was not delivered to Sagane until some six weeks later. (Fletcher Knebel and Charles W. Bailey II, *No High Ground* [New York: Harper and Brothers, 1960], pp. 222–23 and p. 241. This episode is also described in Arthur H. Compton, *Atomic Quest, A Personal Narrative* [New York: Oxford University Press, 1956], p. 259.)

celerator (MTA). Inspired by fears that the Korean War might result in the United States being cut off from foreign supplies of uranium, MTA was essentially an accelerator intended for the purpose of regenerating depleted uranium into weapons-grade plutonium. Panofsky, who served as chief engineer on the project, says that MTA was intended as no more than insurance against the fear (it turned out to be an unfounded fear) that the country was running out of fissionable materials. Though MTA was never needed, Panofsky recalls, it worked, and was later employed for a few years as a basic research instrument. The total cost, he says, was about $40 or $50 million, but when compared with some other abandoned weaponry programs, such as the Skybolt missile, Panofsky notes, "it was not a major fiasco." *

In any case, by 1957, when the SLAC proposal was submitted for federal support, young Panofsky was a rising and highly respected star in the substance of science as well as in the military-oriented scientific councils that had grown up in Washington. He was an adviser to the AEC's Division of Military Applications, as well as a member of the Air Force Scientific Advisory Board, and by 1958, when nuclear test ban talks were proceeding with the Soviets, his wartime experience with devices for detecting nuclear effects led to appointment to the Defense Department's *ad hoc* group on Detection of Nuclear Explosions; a year later, he chaired the United States technical working group that went to Geneva to attempt to negotiate a high altitude test ban. At about the same time, he also became a member of the President's Science Advisory Committee.

This, then, was the man whose proposal for a $100-million accelerator was up for consideration at a time when the long-laboring MURA group had to concede that its designs were still not ready. The

---

* Since basic researchers, in their roles as advisers to the military, have frequently witnessed vast financial debacles involving military hardware, it is not surprising that they view with some disdain the close accountability to which the politicians held the basic research community's relatively tiny share of the federal budget. For example, the Defense Department estimates that between 1943 and 1966, $4.8 billion was spent on various missile systems that were abandoned while in progress or upon completion. Deputy Secretary of Defense Cyrus R. Vance also acknowledged that the Pentagon had "traditionally underestimated the cost of weapons systems very substantially." The total cost of weapons systems abandoned before completion or because of obsolescence after some service was placed at $19.3 billion. (*New York Times*, May 14, 1967.) The difference between defense and basic research, of course, is that defense is readily acknowledged to be a wasteful business, one in which it is safest to err on the side of greater expenditures. Basic research, on the other hand, is more or less incomprehensible to the public and, by definition, remote from utilitarian application.

proposal, it must be recalled, came forth when high energy physicists themselves were beginning to feel concerned about the durability of political support for their ever-costlier field, and, as a consequence, were just starting to talk in terms of choices among various possibilities. Previously, the attitude had been that whatever could be done should be done. Now, however, there was a new note in the discussions: The statesmen of high energy physics were not simply talking of MURA *and* SLAC. They were beginning to speak of MURA *or* SLAC. But Panofsky was determined to proceed with his gigantic machine. Furthermore, he was determined to have it built at Stanford and he was fully up to some brilliant maneuvering to get his way.

With Brookhaven already at work building its giant new machine, Argonne about to break ground, and MURA still bogged down in unresolved design difficulties, a carefully staged series of maneuvers got under way in the summer of 1958 to clear the way for Panofsky to get his machine. At that time, it should be recalled, the White House was newly equipped with the full-time science advisory apparatus that had been installed as a consequence of the panic induced by Sputnik. James R. Killian, Jr., had taken leave of the presidency of MIT to become the first full-time science adviser to the President, and the Science Advisory Committee, which previously had been loosely appended to the White House as part of the Office of Defense Mobilization, was elevated to the status of the President's Science Advisory Committee (PSAC). Thus, for the first time the Executive possessed an effective apparatus for science policy planning. Now, with extraordinary *savoir faire* the high energy community proceeded to press that apparatus into its own service. Killian, a wise, dedicated, but overburdened administrator, thought it would be useful to have NSF reconvene its high energy panel for still another review, immediately to be followed by a joint White House–AEC review that would take the narrow concerns of the high energy physicists and consider them in a national context. Accordingly, in August, 1958, NSF once again assembled its high energy panel.* With the discovery of Dubna having paid off with the

* The membership, now shifted to give greater representation to the Midwest, consisted of Herbert L. Anderson, University of Chicago; Horace R. Crane, University of Michigan; Bernard T. Feld, MIT; Edward J. Lofgren, University of California; Leonard I. Schiff, Stanford; Frederick Seitz, Illinois; Robert Serber, Columbia; M. G. White, Princeton; Robert R. Wilson, Cornell; and Leland J. Haworth, Brookhaven.

ZGS at Argonne, the committee now pointed out that, in the Soviet Union, "a site has been chosen for a 50-bev. alternating gradient pulsed synchrotron." Two years earlier, the panel's predecessor had concluded that American high energy expenditures should rise from the then current $40 million a year to $60 million to $90 million a year by 1962. The 1958 panel stated, however, that "reasonable advances in the program in the next five years would bring the total rate to a level in excess of $100 million and probably to the neighborhood of $150 million per year." As for the proposed Stanford machine, the panel stated that "the proposal is scientifically and technically sound and recommends that this accelerator be built as soon as possible." MURA, though still not ready to move to the construction stage, should be kept alive, the panel recommended, to continue on the FFAG design.[17]

Almost immediately after the completion of the review by the NSF panel, Killian convened a panel to examine the matter "at the Presidential level." In fact, however, the membership of this second panel was drawn from the ranks of high energy practitioners or admiring onlookers, though, in this case, their appointments emanated jointly from PSAC and the General Advisory Committee of the AEC. The chairman was Emanuel R. Piore (later to be the Mohole troubleshooter), a PSAC member, who, after leaving ONR in 1955, later went on to become director of research for IBM; Hans A. Bethe, of Cornell; Leland J. Haworth, director of Brookhaven, Jesse W. Beams, chairman of the physics department at the University of Virginia, and Edwin M. McMillan, associate director (and soon to become director) of the Radiation Laboratory at Berkeley. Convening immediately after the issuance of the NSF report, the Piore panel stated that the need for another review was based in part, on "the extraordinarily high cost of the construction and utilization of high energy accelerators," as well as "the desire to have an orderly national program taking full account of activities abroad." It then proceeded to make an extraordinary statement, apparently based on the assumption that the science-government relationship had evolved to the point where basic science had been given a blank check on the federal treasury. Stated the Piore panel:

"It is not possible to assign relative priorities to various fields of basic science nor should they be placed in competition. Each science, at any given time, faces a set of critical problems that require solutions for continued growth. Sometimes these solutions can be acquired at little cost; sometimes larger expenditures of funds are needed. Hence, the cost may not reflect the relative value but rather the need.

"Each area must be funded according to these needs . . ." [18]

What were the needs of high energy physics? The immediately preceding panel had concluded that between $100 million and $150 million would be required annually by 1963; the Piore panel split the difference and came out for $125 million. The SLAC proposal, it recommended, should be accepted at once "in order to avoid an undue delay in high energy research."

As for MURA, the high energy community was now puckering up for what was eventually to be the kiss of death. The Piore panel recommended that the federal government "reject the present MURA accelerator proposal but continue to provide adequate support and encouragement to the MURA development group." The group, it said, was the source of many valuable ideas in accelerator design, but its designs had not yet reached the point where it would be prudent to begin construction. Thus the more than one hundred people who had been dedicating years of their career to MURA were again left in limbo, though, as always, until the end, a limbo with a note of hope. As McMillan later explained, "The MURA people were all our friends. No one likes to deliver a stroke of death, so we didn't kill it off."

While MURA wallowed in uncertainty, the prospects for SLAC grew ever brighter. It is important to note that Panofsky was promoting a design that had earned the endorsement of the high energy community —including experts drawn from MURA. But perhaps even more important than the design was the presence of Panofsky behind it. As Alvarez said of his former associate, "I backed SLAC because I had faith in the people. I remember that after the war, when Lawrence offered me a choice of any five people to work with, I picked Panofsky." The MURA group was not without distinction, but similar words of unquestioned confidence were not to be found for any of its members.

Thus, the merits of Panofsky as much as the merits of his machine were creating a powerful consensus in the councils that link science and government. On May 14, 1959, this consensus was formally ratified as the policy of the United States government when President Dwight D. Eisenhower formally announced that he would ask the Congress to provide approximately $100 million to build the Stanford accelerator. Speaking at a Symposium on Basic Research sponsored by the National Academy of Sciences, the American Association for the Advancement of Science, and the Alfred P. Sloan Foundation, the President stated, "Physicists consider the project . . . to be of vital importance. Moreover, they believe it promises to make valuable contributions to our understanding in a field in which the United States is already strong, and in which we must maintain our progress." [19] *

Both the President of the United States and the leaders of the high energy physics community now stood behind Panofsky's proposal, and, as events had proceeded up to that point in relations between science and government, this unanimity left no reason to doubt that the required $100 million would at once be forthcoming for construction of the two-mile-long machine. Such, however, was not to be the case, for within days after Eisenhower's speech the SLAC project slowly began to slip into a mesh of political controversy that was to immobilize it for three years. SLAC eventually emerged unscathed, but the same cannot be said for an innocent bystander to these proceedings, MURA.

The fundamental political difficulty of the SLAC proposal was that any new budgetary item of such proportions—at $100 million, it was, up to that time, the costliest single basic research facility ever proposed —was bound to produce a sort of immunological reaction in the delicately strung-together system that controls the disposition of public resources. It may be speculated that if SLAC had not evoked one adverse reaction, it would have evoked another, but in any case, Eisenhower had no sooner delivered his request for Congress to authorize the expenditure of the needed $100 million, than a violent reaction came forth from that unique apotheosis of congressional power, the Joint Committee on Atomic Energy (JCAE). Created by the Atomic Energy Act of 1946, and composed of members drawn from both

---

* Inexorably, the financial appetites of science and technology made folly of Eisenhower's rhetoric of federal frugality. Having come to office on a pledge of cutting federal spending, he helplessly stood by while Congress, in alliance with enthusiasts for health research, increased the NIH budget from $71 million in 1954 to $547 million in the year he left office. In virtually each of these years, the congressional appropriation far exceeded the President's budgetary request. In announcing SLAC he dutifully warned, "Too often we have tended to look unduly to the federal government for initiative and support in a multitude of activities, among them scientific research. . . . Too much dependence upon the federal government may be easy, but too long practiced it can become a dangerous habit." He then proceeded to announce, in reference to SLAC, "Because of the cost, such a project must become a federal responsibility."

houses, the JCAE symbolized congressional determination to establish the legislative branch firmly in the center of policy making for the new and little understood world of the atom. Accordingly, the Congress equipped the JCAE with unprecedented powers, including the requirement that the Atomic Energy Commission "shall keep the Joint Committee fully and currently informed with respect to the Commission's activities." * In 1954, this was amended to read "*all* of the Commission's activities"—as well as matters related to atomic activity under the jurisdiction of the Defense Department.[20] Thus was laid the basis for exempting the affairs of atomic energy from the protective cloak of the executive privilege principle which executive agencies find so useful in their dealings with Congress. Since a congressional committee is no more important than the operating agency whose affairs it commands, it is a common, though not invariable, practice for committees to seek to promote the expansion of the jurisdictions they command. Such being the case, in the midst of the thrift-talking Eisenhower administration, it was the Democratically controlled JCAE that became the focal point of political pressure for a vast expansion of government support for research and development in all fields of atomic energy—from basic research in university laboratories to a decade-long billion-dollar debacle aimed at building an atomic-powered aircraft.† And all along the way, as it promoted an expanded research and development program in atomic energy, the Committee increasingly insisted upon its preeminence in government policymaking on the atom, not only in terms

* Having made the Joint Committee strong, the Congress attended to the other side of the coin by making the Atomic Energy Commission weak. It is the only major agency of government that is headed by a commission—five coequal members—rather than a single director.

† Upon coming to the presidency, one of Kennedy's first moves was to kill this project. It had long before become plain that it was bogged down in vast and costly technical problems, and that even if carried to a successful conclusion, there were serious questions as to what use might be made of such an aircraft. With its radioactive power plant, it certainly could not be permitted to fly over heavily populated areas. Since the Joint Committee and the Eisenhower administration had been increasingly at odds over government support of atomic energy activities, the Committee was poised to fight off efforts to curtail any ongoing atomic research activities, despite the insistence of Defense officials that they did not want the airplane. Defense Secretary Wilson compared it to a "shitepoke—a great big bird that flies over the marshes, that doesn't have too much body or speed to it, or anything, but can fly." (*Aircraft Nuclear Propulsion Program Hearings*, JCAE, 1959, p. 20.) Thus, the atomic plane project lived on under Eisenhower, but with a Democratic president in office, the Committee relented and permitted it to be brought to an end.

of its being the prime mover in initiating projects but also in terms of being minutely informed about the internal affairs of the Atomic Energy Commission.*

On July 14, 1959, exactly three months after Eisenhower announced that his administration had decided to go ahead with the project, the Joint Committee opened a series of hearings titled "The Stanford Linear Electron Accelerator." Representative Melvin Price (D-Ill.), chairman of the committee's subcommittee on research and development, demanded of John A. McCone, chairman of the AEC, "The question arises why the Joint Committee was not informed of the decision to go ahead with this until one day prior to the time the bill was sent up to committee." [21] McCone explained that the decision had been bogged down in lengthy and complex deliberations involving the White House science office, the Bureau of the Budget, the AEC, the Defense Department, and various other nooks and crannies of the federal establishment. Budgets, jurisdictions, policies, and other matters first had to be settled before the AEC could proceed, he explained. "It was because of the necessity of clearing these points, plus the necessity of having the Stanford people come back for final discussions, and to have a final hearing with the Bureau of the Budget . . . that we felt we were not in a position to finally advise the Joint Committee." [22]

Just what prompted the sudden decision to proceed with a proposal that, in fact, had been hanging fire since 1957 did not surface in the public hearings. But Panofsky, who served as an adviser to Killian on arms controls matters, had been strenuously campaigning for a decision to go ahead on the Stanford machine.† Killian, the first full-time White House science adviser, had been gaining in Eisenhower's confidence, and Eisenhower, though wedded to the rhetoric of individualism, was at last coming around to the realization that there was no available

---

* The unique relationship between the JCAE and the Atomic Energy Commission, and the implications of this relationship for public policy making, are carefully described and analyzed in Harold P. Green and Alan Rosenthal, *Government of the Atom* (New York: Atherton Press, 1963). Harold Orlans, *Contracting for Atoms* (Washington: The Brookings Institution, 1967) is a valuable, up-to-date inquiry into the public policy implications of the AEC's methods of contracting.

† A scientist who later came into the picture as a presidential science adviser explains that, in the deliberations leading up to Eisenhower's endorsement of SLAC, Panofsky's greatest strength lay in his scientific reputation and the quality of the SLAC design. But, added this adviser, "It didn't hurt him to be on PSAC."

alternative to large-scale government support for basic research. Killian states that Eisenhower "was always pretty much impressed by the fact that the scientific community was always, in his judgment, seeking greater support, more funds, and he didn't see any way in which decisions could be made on any objective basis as to what the budget should be. He kept asking when we were arguing for more basic research, more research funds, 'How can we tell? Give me some basis of policy to deal with these problems.' " It would be difficult to demonstrate that in 1959, or even today, anything resembling a "policy" has evolved for determining the level of government support for basic research, but the existence of debate and discussion, in contrast to the closed-door decision-making that had prevailed with earlier accelerators, at least created an appearance of SLAC being subjected to the traditional give and take of the public process.

But now a grim irony took command of the politics of high energy physics. Having been nurtured on struggle against the Eisenhower administration's attempts—unsuccessful though they were—to penny-pinch on research, the Joint Committee was now faced with a request to authorize the expenditure of $100 million for a new venture in nuclear research. It is not unlikely that if affairs had been so maneuvered as to make it appear that the initiative for SLAC had come from the JCAE, the Committee would have become the promoter and guardian of the machine, fully prepared to convey it past the administration's budgeteers. But, against the background of long hostility between the JCAE and the White House, and the administration's seeming insensitivity to the Committee's jealous demands to be kept fully informed of AEC affairs, the JCAE declined to move quickly. SLAC came to be referred to as a "Republican accelerator," and the Democrats who controlled the JCAE refused to endorse it. As a consequence, after having promoted breakneck speed in the expansion of nuclear research for more than a decade, the Committee now settled down to several years of examining virtually every aspect of the SLAC proposal, ranging from patent rights on the klystron tubes for the machine (the Committee insisted that Varian Associates be excluded from the project because of the firm's close ties with Stanford) to financing the landscaping of the accelerator site.

Confronted by this unanticipated development, the high energy community responded in traditional fashion: it reconvened the Piore panel. Meeting late in 1960 to set forth the community's wants just prior to the beginning of the Kennedy administration, the panel expressed "the importance of proceeding forthwith" on the SLAC project. "It is important," it stated, "that this accelerator be authorized immediately." [23] With SLAC hanging fire, MURA received no more than an indirect reference. ". . . It is now uncertain," the panel concluded, "whether future programs should include . . . a very high intensity proton accelerator . . . or whether the need can be more usefully met by extension of the energy parameter." Which was a roundabout way of saying that the high-intensity features of the MURA design might be incorporated into a new high energy machine. As for costs, the panel set forth the figure of $350 million to $400 million a year by 1970.

The 1960 Piore report was also the occasion for one of the great rarities of science policy making—a public dissent. Eugene P. Wigner, a distinguished Princeton physicist serving on the panel, attached to the report what he described as a "separate statement," asserting that "the present writer considers the emphasis of the majority report on the significance of high energy physics exaggerated and indeed some statements may even be misleading." Wigner argued that the report failed to consider the effects that an expansion of high energy physics research might have on the availability of financial support and trained scientific manpower for other fields of research. And he added, "we do not expect at present that high energy physics will lead to 'practical results' in the near future. Its aim is a deeper understanding of nature, not the increase of our power to accomplish something tangible." In forwarding the report to the White House, Piore noted in an accompanying letter that the "Panel did not study and has not attempted to evaluate either the possible effects on other branches of science or the ability of the country to support the program. The conclusions reached are consistent with the idea of increasing emphasis on all branches of basic science." [24] Wigner, in effect, was arguing for what is anathema to the ideology of pure science—an ordering of priorities; Piore, on the other hand, was holding to the traditional view that science was undersupported and underdeveloped, and that all fields merited expansion without favoritism.

Still, the Joint Committee refused to budge, for a new reason arose for holding back on the SLAC project. At Hanford, Washington, where the AEC had built an immense facility for the production of plutonium for atomic weapons, a new reactor was scheduled to be constructed. In the existing plant, the immense heat that came out as a by-product of plutonium production was simply carried off by running the waters of the Columbia River through the reactors. Rather than continue to waste this heat, the Committee reasoned, why not harness it for the production of electric power? Technically and economically the proposal made sense, but the coal industry, already depressed by expanded use of gas and oil, rebelled at the creation of a precedent for producing power with government-subsidized atomic energy. And the Eisenhower administration, dedicated to a cautiously paced development of atomic energy for civilian purposes, similarly rebelled at the prospect of putting the AEC into the power business. Thus arose the euphonious deadlock of Hanford and Stanford. Clearly, there was no substantive connection between Panofsky's long-stalled accelerator project and the Joint Committee's determination to draw steam for electric power out of the new reactor planned at Hanford. Nevertheless, since Eisenhower refused to approve the Committee's scheme for Hanford, the Committee refused to approve his scheme for Stanford. In April 1960, six of the Committee's eight Republicans vainly fought to restore SLAC to the budget, but when the issue came to the floor of the House, the accelerator was defeated in a vote that closely paralleled party lines. The Democrats went against it 188 to 10; the Republicans supported it 118 to 7. At that point, it could be said that if Panofsky's machine could split atoms as effectively as it split the House, it would indeed be a success.

For purposes of exploring the politics of science, there is no point in going into all of the bizarre proceedings that followed the irrelevant linkage of Hanford and Stanford. Perhaps one instance, however, will suffice to convey the flavor. When it became known that Panofsky's design called for the construction of a tunnel to house the two-mile machine, several legislators from the State of Washington came forward to volunteer an abandoned railroad tunnel near Stevens Pass, Washington. Since Panofsky had no desire to locate his machine away from the pleasant environs of Palo Alto, California, and the AEC felt that what Panofsky wants Panofsky should get, he and the Commission proceeded to lay down a barrage of argument against the Washington proposal, faulting it on every possible score. Several years later, when the machine was nearing completion—at Stanford—a SLAC scientist laughingly pointed out that "the AEC rejected the [Washington] tunnel on the grounds that it had a slope of one-quarter of one percent." Pointing to the vast machine being built at Stanford, he added, "The tunnel we've built here has a slope of one-half of one percent." At issue, of course, was not the slope; rather it was, as the scientist put it, "locating the machine in a place where an intellectually viable group could thrive." High energy physicists are perhaps the most urbane members of the scientific community, and they do not take well to living in the provinces.

With a Democratic administration in office, relative harmony came to prevail between the Joint Committee and the White House. In 1961, the go-ahead was given for the Stanford project, and the following year brought approval for a scaled-down, though still sizable, Hanford power project. But where did this leave MURA? And even more important, where did this leave the rapidly advancing plans for still a new machine, of at least 100 bev., that was being designed at the particle research center that bore the name of the great pioneer of the field, the Lawrence Radiation Laboratory, near the Berkeley campus of the University of California? These plans were near the point where the actual construction work could soon begin. Once again, a committee was convened to formulate a recommendation.

An ever-thickening mist of nostalgia distorts our vision of the short-lived Kennedy administration. I concede that I may be a victim of this nostalgia. But, as a working journalist, covering the affairs of science and government during this period, my recollection is that Kennedy had a peculiar infatuation for the ways and men of science. And nowhere was this more manifest than in his relationship with his science adviser, Jerome B. Wiesner. Wiesner, in an elegant memorial written on the day of the President's burial, observed of Kennedy, "His background ill prepared him for an interest in scientific matters, yet his interest was lively. . . . For example, he was forever trying to get someone to explain electromagnetic propagation comprehensibly. . . . He wanted to know how radio worked. But when one tried to answer, one learned that the question was not about electron tubes or transisters or cells—these were manmade things which he could believe—but why and how did nature really allow energy to be sent through space." [25]

Kennedy's interest in science was indeed lively, but coupled with that interest was a realization that, in the early 1960's, after fifteen years of fantastically rapid growth of federal support for basic research, the Congress was feeling sour and restless about the seemingly endless financial appetite of the nation's scientific community. After several discussions with Wiesner on the most basic and most expensive of the sciences—high energy physics—he informally gave an assurance that his administration would back the construction of one major new accelerator about every five years—but that more than that was not likely to be had for the foreseeable future. It was against this background—with SLAC safely past the Congress, MURA still pending, and the Berkeley proposal ripe for consideration—that a panel of scientific statesmen was assembled, in November 1962, to select the next machine to be constructed. This was to be a summit conference of high energy physics, or, to be more precise, a Munich conference, with the Bureau of the Budget as the power to be appeased and MURA in the role of Czechoslovakia.

With the same attention to geographic origin that is traditionally accorded vice-presidential candidates, the new panel contained three from the East Coast: Chairman Norman F. Ramsey, of Harvard, E. M. Purcell, also of Harvard, and T. D. Lee, of Columbia. Three from the Midwest: E. L. Goldwasser, of the University of Illinois; Frederick Seitz, who was then dividing his time between Illinois and the presidency of the National Academy of Sciences, and John H. Williams, of the University of Minnesota. And three from the Far West: Owen Chamberlain, of the University of California, Murray Gell-Mann, of Caltech, and Panofsky, of Stanford. The tenth and final member of the panel, Philip H. Abelson, editor of *Science* and director of the Carnegie Geophysical Laboratory, was situated in Washington, had switched from physics to geophysics, and was in both geography and discipline remote from high energy physics. Serving as consultants to what came to be referred to as the Ramsey panel were sixty-five specialists of varying sorts—high energy physicists drawn from every major machine in the country, two staff men from the Bureau of the Budget, several of NSF's experts on the training and availability of scientific manpower, a representative from the space agency and the director and three staff members of MURA. Rabi, the behind-the-scenes statesman, was not on the panel or among its listed consultants. But Ramsey was a former student of the influential Rabi, Purcell had admiringly worked under him at the MIT Rad Lab during World War II, and Lee was a colleague of his on the Columbia physics faculty. And they all knew Rabi's viewpoint, which was that high energy physics had got itself into a political bind by thinking too small. It was Rabi's opinion that many problems would never have appeared if the federal government had spread around the country some half-dozen replicas of the vast research center he had helped to build at Brookhaven. Unfortunately for this point of view, the time had passed when high energy physics had easy entree to the federal treasury.

In 1958, when the Piore panel wrote its first report, high energy physics was tone-deaf to discussions of economy. In 1960, when the panel wrote its second report, Wigner's dissent caused his colleagues to regard him as the skunk at the picnic. But by 1962, the same tightening of money that was so disconcerting to the chemists was leading the high energy physicists to the realization—at Wiesner's insistence, per Kenndey's instructions—that it would be imprudent even to *ask* for everything they wanted. Accordingly, the report of the Ramsey panel opened with the assurance that "Although the [recommended construction] program is obviously expensive, it is at the same time restrained in the sense that it is limited and selective in the number of new facilities to be provided." [26] Having proclaimed selectivity as a feature of its findings, the panel proceeded to state that "energy is the single most important parameter to be extended," which meant that the MURA design, which was distinguished for intensity rather than energy, did not rate top priority. Then the panel proceeded to narrow the issue further by declaring the necessity of maintaining "the continuing productivity of the major U. S. high-energy laboratories, in particular, the Brookhaven National Laboratory and the Lawrence Radiation Laboratory." Under a heading of "Discussion of Some Principal Recommendations," which preceded the actual recommendations, the report stated that the laboratories at Brookhaven and Berkeley "are the principal source of U. S. leadership in high energy physics." Then it went on to say, "The Panel believes that its recommendations will preserve the vitality of these centers on the east and west coast while at the same time strengthening the contribution from the Middle West through its recommendation of proceeding with the construction of the MURA accelerator."

When it came to the actual recommendations, was the panel actually endorsing the construction of the MURA accelerator? An impression to this effect is certainly suggested by the passage just quoted, but, in fact, the panel, confronted by a tightening budget and a determination to pursue higher energies, was willing to throttle MURA. The problem at hand was how to do it without leaving fingerprints.*

With higher energies unanimously agreed upon as the most urgent need for the progress of the high energy physics, the Ramsey panel recommended "construction *by* the Lawrence Radiation Laboratory of a high-energy proton accelerator of approximately 200-bev. energy . . ." (Italics supplied, for the interpretation of *by* was itself to become the focus of a vast struggle a few years later.) Next, the panel proposed an extensive construction program to raise the power of the 33-bev. machine at Brookhaven, and, third, on the subject of higher energies, it recommended that Brookhaven be given support to design a 600- to 1000-bev. machine, with the start of construction planned for around the end of the decade. The cost of building this machine was estimated at approximately $800 million. Was any order of priority prescribed in this three-part program? Not quite, for the report stated: "It should be noted that step 3 . . . is recommended with equal urgency as step 1." But what was to be done with MURA? The answer was that its design should be constructed, "provided this is not expected to delay significantly the authorization of the steps toward higher energy recommended above." Having delivered what it full well knew to be lethal faint praise, the report then went on to excessive protestation: "The panel considered recommending against the construction of the FFAG accelerator," it explained, but it was convinced that there was useful work to be done with the type of machine MURA had designed. Nevertheless, the panel explained, higher energy was more desirable than higher intensity. It might be thought that this sufficed to state the panel's conclusions, but, the panel well knew that it had sealed the doom of

---

* It can be asked how this could be accomplished when three midwesterners sat on the panel. The answer is that their loyalty to the progress of high energy physics outweighed their affection for the Midwest. Even among MURA's most ardent partisans, there was little doubt that, if a choice had to be made, scientific judgment called for putting energy ahead of intensity. However, MURA's contention was that both machines were of great scientific importance, and should not be regarded as mutually exclusive. Further contributing to the Ramsey panel's unanimity was a general aversion to having scientific disputes come into public view. Abelson, for example, voted with the panel, resigned, and later criticized the panel's budgetary recommendations on the grounds that high energy physics was seeking a disproportionately large share of federal research expenditures. (See Abelson's testimony, *High Energy Physics Research, Hearings,* JCAE, March 1965, pp. 209 ff.)

MURA when it assigned secondary importance to higher intensity, and since it felt troubled by what it was doing to people who had been laboring on hope for nearly a decade, the writers of the Ramsey panel report chose to state the matter all over again: "The panel concluded that it should state clearly its views in this matter; namely, that the highest priority in new accelerator construction should be assigned to the recommended steps toward highest attainable energy, but that a 12.5-bev. high intensity FFAG accelerator is an essential component of a balanced program and should be constructed provided that it will not delay the authorization of the steps toward higher energy."

The faintheartedness of the MURA endorsement was glaringly illuminated when the panel next proceeded to endorse, without qualification, the construction of a 10-bev. machine at Cornell. Though little attention had been evoked by this machine, it was the pet project of one of the most professionally admired figures in the inner core of high energy physics, Robert R. Wilson, who, during World War II, had headed the experimental nuclear physics division at Los Alamos. As one White House science adviser put it, "Wilson was very persuasive." This adviser added that, though he did not consider Ithaca a suitable site for a federally financed facility that was theoretically supposed to serve scientists throughout the nation, Wilson's reputation as an accelerator builder was such that his colleagues were reluctant to turn him down. It was to Wilson's advantage that his machine was relatively modest in price—approximately $10 million—and that NSF was willing to pay for it, rather than the cost being imposed on the hard-pressed high energy physics budget of the AEC. It might be added that, in line with the well-established pattern of interlocking positions in the leadership of the scientific community, Wilson, the recipient of NSF largesse, also served as a member of one of NSF's highest ranking advisory bodies, the division committee for the mathematical, physical, and engineering sciences. He got his machine.

Though the Ramsey panel introduced its report with a claim of being "limited and selective" in recommending new facilities, the overall price estimate for its designs came to a total of more than $8.2 billion over twenty years. This figure was staggering by itself, but among those who knew the business, there was the realization that up to that point, even the most generous cost estimates for accelerator construction had invariably turned out to be grossly inadequate.* High energy physics, theretofore the most generously supported field of research in the post-war period, was now on the brink of entering into a scarcity economy. The weakest participant in this economy was the MURA group, which at that point, consisted of forty scientists and engineers, plus seventy technical and administrative employees. Having been rejected within the scientific community, MURA now proceeded to fight for its life by going outside of science politics and into national politics. In the field of military technology, there was ample precedent for scientists to seek alliances with politicians in efforts to have their weapons concepts incorporated into strategic doctrine. Edward Teller and his colleagues had done just that in their campaign to commit the AEC to building a hydrogen bomb. But never before had the esoteric implements of pure science been made the subject of controversy within the arena of traditional politics. MURA, however, had nowhere else to turn. A West Coast Nobel laureate recalls that the subject of MURA's plight came up in, of all places, Dubna, while he and several MURA scientists were members of a delegation visiting the Soviet high energy physics center. He reports that over drinks in a hotel, "I told them that it was a great mistake to take the political route, and they said, 'That's all we can do, we don't have any Nobel Prize winners on our side.' "

NOTES

1. Johnson, letter to Senator Hubert Humphrey, January 16, 1964.
2. *Report of the Panel on High Energy Accelerator Physics,* April 26, 1963, Division of Technical Information, Atomic Energy Commission, p. 42.
3. *High Energy Physics Program,* report on National Policy and Background Information, Joint Committee on Atomic Energy, 1965, p. 141.
4. R. G. Hewlett and O. E. Anderson, *The New World, 1939/46,* Vol. I, *A History of the United States Atomic Energy Commission* (University Park, Pa.: Pennsylvania State University Press, 1962), p. 635.
5. Quoted in *High Energy Physics Research,* Joint Committee on Atomic Energy, 1965, p. 189.

* In 1961, Chairman Holifield, of the Joint Committee, stated, "I do not think we have ever built any one of these accelerators . . . yet that didn't run two or three times the original estimate." Replied AEC Chairman John McCone, "I guess that is right." (*AEC Authorizing Legislation,* JCAE, 1961, p. 89.) When SLAC was completed, in 1967, it set a precedent by being almost precisely within the original cost estimate for construction, but, as it turned out, the operating costs had been markedly underestimated.

6. *Ibid.*
7. *Ibid.,* p. 190.
8. Report of the NSF Advisory Panel on Ultrahigh Energy Nuclear Accelerators, May 2, 1954, *High Energy Physics Program* (1965), p. 165.
9. Report of the Advisory Panel on High Energy Accelerators to the National Science Foundation, October 25, 1956, *High Energy Physics Program* (1965).
10. AEC Press Release No. 785.
11. *AEC Authorizing Legislation,* 1956, Joint Committee on Atomic Energy.
12. *Ibid.,* 1957, p. 63.
13. *Ibid.,* p. 61.
14. *AEC Authorizing Legislation,* 1958, pp. 183–84.
15. Quoted in *Stanford Linear Electron Accelerator, Hearings,* JCAE, 1959, p. 60.
16. *AEC Authorizing Legislation,* 1958, p. 194.
17. Supplement to the Report of the Advisory Panel on High Energy Accelerators to the National Science Foundation, August 7–8, 1958, *High Energy Physics Program,* pp. 143–47.
18. Piore Panel Report, 1958, *High Energy Physics Program,* p. 138.
19. "Science: Handmaiden of Freedom," address to Symposium on Basic Research, Publication No. 56, American Association for Advancement of Science, 1959, pp. 133–42.
20. 68 Stat. 956 (1954); 42 U.S.C. 2252 (1958).
21. *Stanford Linear Electron Accelerator, Hearings,* JCAE, 1959, p. 11.
22. *Ibid.,* p. 12.
23. Piore Panel Report, 1960, *High Energy Physics Program,* p. 127.
24. Background Information on the High Energy Physics Program and the Proposed Stanford Linear Electron Accelerator Project, Report to the Joint Committee on Atomic Energy, 1961, p. 102.
25. Jerome B. Wiesner, "John F. Kennedy: A Remembrance," *Science,* November 29, 1963.
26. *High Energy Physics Program,* p. 86.

XI

MURA's Last Stand

The failure to approve an accelerator for the Midwest would seriously compromise the prospects for approving a $250-million accelerator on the east or west coasts. . . . I say this not with any notion that there might be some kind of political retaliation. I say this from the standpoint of realism.

SENATOR WILLIAM PROXMIRE (D-Wis.)
to Jerome B. Wiesner, White House
Science Adviser, 1963

For the scientists and engineers on MURA's staff, the task of designing the FFAG accelerator was not simply a source of employment. Many of these people were among the most skilled and sought-after in their esoteric craft. And at a time of booming opportunities in academic and industrial research, they had chosen to remain with MURA because they had faith that, eventually, their difficult and long-gestating design would be translated into a gigantic research apparatus—at a cost of $170 million to the AEC. For many of them, the FFAG design, which had evolved over a decade, was the principal tangible product of their professional careers. Rather than stand by peaceably while the Ramsey panel, with its preference for higher energies, sacrificed the MURA machine as evidence of fiscal responsibility, they were prepared to fight. And, as it turned out, the politicians of the Midwest were ready to fight for them.

It must not be thought, however, that these new and combat-prone allies of the midwestern scientists were inspired by sentiments, one way or the other, on the crucial issue of particle intensity versus particle energy. The arcane substance of the MURA controversy did not lend itself to lay understanding, but the economics of the matter were flawlessly lucid to even the most dim-sighted politician: In the second decade of the science-government relationship, research and development had become the new public pork barrel, and though few politicians had a comprehension of what high energy physics was about (in fact, it was, and is, not unusual for congressmen to refer to accelerators as "reactors") they knew that it was expensive, and they knew that glory and power were to be had for those who could bring such federal monuments home to their constituents. Furthermore, this pork barrel instinct, though prevalent in every part of the country, was particularly intense in the Midwest. For that region, with its economy based on heavy industry, consumer goods, and agriculture, was coming to the recognition that while it produced steel, automobiles, and dairy products, other regions of the country, especially the Northeast and the West Coast, were assuring their own future prosperity by becoming the principal producers of the federally subsidized advanced technologies that loomed ever larger in the national economy. In statistical studies that were evoked by the sense of being left out, it was found that California, with its heavy concentrations of aerospace and electronics industries and research-oriented universities, was receiving 34.6 percent of *all* federal expenditures for research and development; the California total, over $4.9 billion, averaged out to $315.14 for each inhabitant of that state. Missouri, mainly because of the presence of McDonnell Aircraft, in St. Louis, led the Midwest, but with only 2.7 percent of the national total, and a per capita figure of $89.42. Massachusetts, with its famous ring of electronics firms on Route 128 around Boston, plus grant-laden Harvard and MIT, received 4 percent of the total, and $110.76 per capita, which was more in total and per capita than Illinois, Michigan, Indiana, and Minnesota combined.[1] It is, of course, possible to slice these numbers every which way and, depending upon the criteria employed, cover a generous spectrum of conclusions as to the justice of the R and D distribution. For example, while the East and West coasts received a vast proportion of federal funds for industrial research, funds for academic research were far more evenly spread throughout the country. Thus, five midwestern universities ranked among the top 15 in receipt of federal research funds in 1965. And though midwesterners regularly singled out Harvard as the symbol of maldistribution, the fact was that Harvard, with $39.3 million in federal receipts for research, ranked behind the University of Michigan ($50.2 million), and the University of Illinois ($40.5 million).[2] Furthermore, while midwesterners might point accusing fingers at Massachusetts or California, the fact was that, within those states, many respectable institutions were quite remote from affluence. Harvard's neighbor, Tufts University, with a mere $5.7 million in federal receipts, knew little of the hold that Massachusetts was purported to have on the federal treasury. Nevertheless, in the very same period that the MURA issue was coming to a climax within the scientific community, the issue of geographic distribution of research and development funds was beginning to ignite passions in the political community. And the passionate politicians were not concerned with statistical subtleties. Thus, in 1962 Minnesota's Senator Hubert H. Humphrey, upon learning that universities and other nonprofit institutions in his home state received $3 million from the Department of Defense, as compared with $120 million for their Massachusetts counterparts, proclaimed on the floor of the Senate, "With all due respect to that great institution, the Massachusetts Institute of Technology, they are just not that good. And the Institute of Technology at the University of Minnesota is just not that relatively unable to do the job." [3] The explanation, if not the justification for the imbalance, was that a major portion of Massachusett's $120 million went to the great military electronics laboratories that MIT founded for the government during World War II and continued to operate after the war under contract to the Defense Department. In any case, the complexities of the science-government relationship were such that charges and accusations concerning the fairness of the distribution of funds were easy to formulate; rebuttals and explanations automatically became enmeshed in statistical and historical peculiarities.

But there was not doubt that if any single rule governed the distribution of federal research and development funds it was simply the biblical rule of Matthew, "Unto every one that hath shall be given, and he shall have abundance. . . ." During World War II, OSRD sensibly followed a policy of favoring the use and, if necessary, the expansion of existing research facilities in preference to building new ones from the ground up. Thus, the surge of wartime research spending helped to reinforce the prewar lead of those areas where, for whatever reason, academic science and technologically advanced industry had already taken root. When war came, Detroit's automotive industry turned from cars and trucks to tanks and jeeps; after the war, it returned to cars and trucks. The West Coast's budding aviation industry flourished when war came, flourished further when peace came, and quite easily made the transition to space work in the 1950's. The universities that were strong in science before the war became stronger during the war. Their strength not only made them attractive to the dispensers of money in Washington, but also influenced the Washington end of the financial system to call upon scientists from these very same universities for counsel on distributing the money. In the view of East and West Coast scientists and industrialists justice had indeed prevailed. But in the view of the self-styled "have-nots," nothing resembling justice prevailed when a single state received one out of every three federal dollars for research and development. Congressmen seeking government contracts for their constituents, and the community booster organizations that prod them on to ever greater efforts to bring home federal largesse, are wholly unfettered when it comes to prescribing systems of just distribution. What it comes down to is that justice and fairness manifest themselves through the safe delivery of contracts to the home district.*

In 1962, a nationwide economic recession inspired many congressmen to intensify their efforts to help acquire government contracts for their constituents. Since the Defense Department accounted for roughly half of all federal expenditures, it became the favorite object of congressional badgering. In response, if not in self-defense, the Department prepared a detailed study, "The Changing Patterns of Defense Procurement," which sought to describe and analyze the processes governing the placement of its massive contracts. Issued over the signature of Deputy Defense Secretary Roswell L. Gilpatric, the study noted that in the years that had passed since the Korean War, the emphasis in military procurement had passed from long production runs of standardized equipment, such as tanks and trucks, to new and often handcrafted advanced technology hardware, such as missile systems and electronic computers. In the fiscal years 1951–1953, Michigan, for example, had received 9.5 percent of the Department's procurement expenditures; by fiscal 1961, the proportion had dropped to 2.7 per-

---

* In the early 1960's, when Houston was vying to be the site of NASA's Manned Spacecraft Center, Houston boosters argued that their region had been denied a fair share of federal research funds. Houston got the Center. Several years later, when it was competing for a nuclear accelerator, the *Houston Post* argued editorially that "the presence of the space center enhances Houston's attractiveness . . . and has helped tremendously to create the sort of community which the laboratory's staff would find congenial and conducive to its work." (*Houston Post*, June 17, 1965.)

cent; in the same period, California's share had risen from 13.6 percent to 23.9 percent. In assessing the genesis of the various technological and industrial capabilities that the Defense Department employed for meeting its research and material requirements, the study concluded, "In summary, successful research and design, or development and testing effort, often leads to follow-on production contracts; and, in turn, engineering work on highly complex new weapon systems creates new R and D capability. The process is circular; and it regenerates itself." [4] But the portion of the study that attracted most congressional attention was a quotation from what was actually a self-serving sales brochure published by the newly founded Graduate Research Center of the Southwest. The Center, which was trying to establish itself as a focal point for academic and industrial research in the Dallas region, stated as gospel the proposition that "Management planners, in considering sites for new or expanded facilities, have found that the availability of trained minds overshadows even such factors as the labor market, water supply, and power sources. The evidence is overwhelming: Route 128 encircling Boston, the industrial complex around San Francisco Bay, that related to the California Institute of Technology and UCLA in the Los Angeles area, and other similar situations are cogent examples of the clustering of industry around centers of learning. Such a migration arises from the need by industry for access to persons with advanced training who can translate the new science into vastly improved or wholly new products." [5]

The Defense Department study pointed out that the quoted analysis was considered to be an "overstatement" by some persons, and that a great deal of defense industry prospered in the absence of "centers of learning," while such centers often prospered without attracting defense industry. The midwestern states constituted a devastating rebuttal to the generalization, for one of the most galling ironies of their place in national scientific affairs was that they were net exporters of scientists and engineers. The great midwestern universities trained them, and then, in large numbers they took up employment with East and West Coast organizations existing almost wholly, if not altogether, on government funds. Thus, it was found, for example, that in nine midwestern states, 43,614 engineering degrees were conferred between 1954 and 1958. But in 1960, only 32,717 engineers were employed in those states.[6] It was estimated that 55 to 75 percent of Iowa's college and university graduates left the state for their first jobs.[7] But, paradoxically, one of the principal lessons drawn from the Gilpatric study was that university science is a magnet for defense contracts. The midwestern congressional delegations, accordingly, became more and more interested in the processes governing the allocation of resources for basic scientific research. The politicians had not suddenly acquired affection for the substance of science, nor did they seek to coat their interest with such a pretense. Rather, they now were proceeding, at least rhetorically, on the assumption that basic science is good for business. As sixteen midwestern senators supporting MURA put it, in a statement that they sent to the White House on August 6, 1963, "Industry which is truly modern flows to the centers of scientific research." Subsequent systematic analysis produced little support for this generalization. Albert Shapero, professor of management at the University of Texas, who conducted an extensive study on factors affecting regional economic development, concluded that the relationship between university science departments and regional economic development was, with the spectacular exceptions of MIT and Stanford, at best quite tenuous. "There are over 1500 four-year-plus colleges and universities in the United States," Shapero noted. "Of these, over 350 have graduate degree programs and yet almost none of them have been credited with company spin-off's." [8] Nevertheless, as the MURA issue approached a boil, the midwestern legislators were becoming fixed on the assumption that basic science was a surefire device for bringing industry to their constituents.

The Ramsey Report, issued on May 20, 1963, left MURA with little time to struggle if funds for the FFAG machine were to be included in the budget that the President would send to Congress early the following year. In the cumbersome workings of the federal budgetary process, the general outlines of each agency's budget had to be settled by midsummer, a fairly precise budget had to be worked out by late fall, and final details had to be settled by mid-December, so that the massive budget volume, exceeding 1000 closely packed pages of figures and text, could be printed and delivered to Congress with the President's annual budget message around the third week in January. Then would begin the laborious and rarely predictable congressional authorization and appropriation processes, which, if all went well, would result in funds being made available for fiscal 1965, which began on July 1, 1964. As things stood then, MURA was having a difficult time holding its staff together on a diet of hope. Following the blow of the Ramsey Report's faint praise, it was clear that if the FFAG machine were not funded for the coming fiscal year, the MURA staff would drift off to more promising employment, and the FFAG design, whatever its scientific virtues, would be consigned to the archives of the AEC. Accordingly, MURA began what it correctly foresaw to be a final fight for survival.

The opening move came on July 8, 1963, when the midwestern organization issued a press release proclaiming, "Scientists from fifteen major universities in the nine Midwest states which train a third of the nation's physical scientists are ready to construct the MURA atom smasher which will make possible studies of 'inner space' far beyond any similar equipment anywhere in the world." Stating that the MURA machine would help utilize "more fully the high energy physics talent in the heart of the nation," and emphasizing that the group was prepared to commence construction *now,* the press release blithely concluded with the assertion, "Federal funds have been requested for construction of the new accelerator and a panel of scientists from the President's Science Advisory Committee and the Atomic Energy Commission's General Advisory Committee has recommended its construction in 1965 as part of the nation's total high energy accelerator physics program. The request and the recommendation now are before the President." The press release contained no inaccuracy. It did, however, fail to explain that the panel of scientists to which it referred had explicitly accorded a secondary priority to the MURA machine, and had tacitly consented to sacrificing MURA in appeasement of the Bureau of the Budget.

While MURA was fighting its case in the public press, its scientists and engineers, most of whom held faculty appointments in the organization's fifteen member universities, were also employing another tactic in the struggle for survival. They were enlisting the prestige of the presidents of their institutions to obtain the support of midwestern elective officials. In this exercise, that besieged and battered breed of executive, the university president, had the uncommon opportunity of standing only to profit—if in no more than faculty goodwill—for venturing into a nasty controversy. As for the politicians, they were newly sensitive to the issue of geographic distribution of research funds and here was the most concrete case that had ever come their way: the midwestern accelerator versus the California accelerator. It was as clear-cut as that. Furthermore, politicians, especially Washington-based politicians, who prefer to remain aloof from the controversies that frequently swirl around state universities, correctly saw no danger in extending themselves on this particular issue. Favorable publicity would accompany their efforts, regardless of the outcome; the university presidents, who were by and large unfamiliar with the ways of doing business in Washington, would be deeply grateful for a show of solicitude by their representatives and Senators. If, at no cost, a politician could have a university in his debt—well, why not? University presidents and trustees are among the most influential members of their communities, and even if they are not for you at election time, memories of assistance rendered the university can diminish, or perhaps even neutralize, their opposition.

By midsummer of 1963, MURA's survival fight was fully under way, and it was to remain at a high pitch until close to the end of the year. In July, at the instigation of MURA's university presidents, congressmen from the Midwest were placing their signatures on form letters that told President John F. Kennedy that MURA "represents only a crucial first step in restoring the Midwest to its rightful place—along with the East and West Coasts—as a major center for educational research. . . . Mr. President, it is crucial that funds for the MURA high intensity accelerator be included in your fiscal 1965 budget." By the dozens, these and similar letters were dispatched to the White House, the Bureau of the Budget, and the Atomic Energy Commission.

In reply, the administration hewed to the line that a decision had not yet been made on the MURA issue, but along with this routine soothing treatment, there began to appear an ominous note. On July 23 Budget Director Kermit Gordon wrote to one petitioning Midwest congressman: "With regard to your reference to high energy physics in the Middle West, I would note as a matter of interest that the ZGS high energy accelerator at the Argonne National Laboratory in Illinois, shortly to be in operation, is the second most costly project of this sort ever to be undertaken in the United States. A major factor in the selection of the Argonne site was that this machine would serve the needs of the midwestern universities, and I am confident that it will be successful in this regard." Gordon's reference to the Argonne machine was based in fact; the ZGS was intended to serve the physicists of the

Midwest. But Panofsky's selection of Stanford as the site for SLAC never encountered the argument that a major accelerator at nearby Berkeley existed to serve physicists in the San Francisco Bay area. Clearly, the midwesterners felt, they were being accorded second-class status by the paymasters of federal funds for research. The congressman who received Gordon's reply forwarded a copy to Indiana University president Elvis J. Stahr, Jr., who served as president of the MURA consortium. In response Stahr urged the congressman on to more intensive efforts in behalf of MURA, noting that "of *ten* high-energy accelerators which are thus far authorized in the United States —operating or under construction—only *one,* that nearing completion at Argonne in Chicago, is located in the Midwest.

"The scientists of the Midwest," he continued, "are thoroughly convinced that this one will not meet their needs. . . . [It] is vital to the future growth of the region and the national economic balance that scientific research by our faculties and for our industries be stepped up relative to the East and West."

For the congressmen who were sending back and forth such correspondence, the MURA issue was a political delight. By championing the machine, they were striking alliances with some of the most influential—and, normally, politically aloof—elements of their constituencies. And, in contrast to the general rules of political alliance, there were no liabilities, risks, or trade-offs in being associated with the campaign to persuade the administration to provide the money for MURA's machine. In fact, the most raucous advocacy of the machine did not even raise the danger of antagonizing the administration. Its legislative liaison men knew that virtually all congressmen hungered and schemed for opportunities to demonstrate to their constituents that they were working for them in Washington. And when the White House and the Congress were ruled by the same party—as was the case in 1963—the heads of executive agencies were pleased to assist their legislative compatriots in conveying the impression of furious and effective activity in behalf of the voters back home. The correspondence between the congressmen and the agency executives flowed back and forth, each round generating a congressional press release to news media back home. The agencies often cooperated to the extent of replying to congressmen's letters in triplicate, so that the legislative champions could send home authentic evidence of labor in behalf of their constituents. As for Kennedy, a dozen years of service on Capitol Hill had educated him to the realities of legislative bombastics. Wiesner recalls that the late President got a bit annoyed when the midwesterners contrived opportunities to lobby at the White House in behalf of the MURA machine, but, by and large, Wiesner says, Kennedy looked upon the situation with amusement. "The pressure was from people that he knew. He was on friendly terms with them. He just looked upon their lobbying as part of the game."

The midwesterners did, indeed, have grounds for revolting against the distribution of federal research funds. But their argument that high energy physics would draw industry to the Midwest had no basis in fact. The world's biggest accelerator, the AGS at Brookhaven, could take no credit for any significant industrial development on Long Island. And though midwestern congressmen happily jumped into the MURA battle, few, if any of them, bothered to recognize that the issue of geographic distribution of research funds was intimately related to the volume of funds available for research. In the same year that they were conspicuously battling for MURA, the House and Senate, with virtually no debate, reduced by $236 million the amount Kennedy sought for the National Science Foundation. None of the midwesterners looked up from the MURA battle to take note of the evisceration of this budget, though, NSF within the bounds of its relatively limited funds, had developed great skill in promoting the development of academic science—which, after all, in the midwesterners' own terms, was what the MURA fight was all about. Nevertheless, in the best style of American political irrationality, MURA became an obsession. Wiesner reports, "I don't think I had as much mail pro and con on the nuclear test ban as I had on the MURA machine."

As the summer of 1963 drew to an end, and the deadline approached for deciding whether the MURA project was to be included in the budget for the coming fiscal year, outward appearances suggested that the Argonne-MURA antagonisms had been smothered beneath the area-wide united front that had been organized in behalf of the Wisconsin accelerator. However, the fact of the matter was that at Argonne, where the nearly completed ZGS would soon give the Midwest its first major high energy facility, there was little joy at the prospect of the ancient enemy, MURA, getting the go ahead to build a machine far costlier and, perhaps, even scientifically more interesting, than the ZGS. Accordingly, while Washington's science administrators and budget makers were deluged by midwestern pleas in behalf of MURA, they were receiving some quite contrary communications from persons associated with Argonne.

The extent and intensity of these efforts at sabotage from within are difficult to ascertain, for after years of intraregional strife over high energy physics, the Midwest was placing a high premium upon harmony and unanimity in behalf of MURA. But some flavor of the sentiments prevailing at Argonne can be obtained from a letter that Arthur Roberts, an Argonne senior staff physicist, wrote to Paul W. McDaniel, director of the AEC's physical research division, on September 6. Stating that he had sampled opinions "from as many physicists as I could," and claiming to "represent a significant fraction of current midwestern opinion," Roberts warned that MURA would prove to be lethal competition for Argonne. The MURA machine—with an estimated construction cost at least three times that of Argonne's ZGS—incorporated and surpassed many of the ZGS's features. Noting that in the Ramsey panel report, "The shutdown of obsolescent accelerators is explicitly advised," Roberts warned that "there would be nothing ZGS could do that FFAG's would not do better.

"The new-born ZGS stands at present on the point of becoming one of the world's most important and productive accelerators," Roberts stated. "We should now be engaged in devising ways of nurturing our baby in every way, not in preparing a coffin for its burial."

Meanwhile, the Chicago press took up the theme that MURA represented a long-range threat to the growth, if not the survival, of Argonne as a high energy research center. Under a headline, "War Over Two Atom Smashers," the Chicago *Sun-Times,* affectionately referring to Argonne's ZGS as "Ziggy," warned that the machine, "in the opinion of responsible scientists at Argonne, is threatened with premature obsolescence and extinction by the plan to build FAG [sic] only 150 miles away. The reason is FAG is designed to do the same kind of job Ziggy does, but to do it 100 times better. FAG will thus make Ziggy obsolete when it is completed in the 1970's." Oblivious of the hundreds of millions of dollars that the AEC had spent to make Argonne a foremost reactor research center, an accompanying editorial asserted, "Until the Argonne machine [the ZGS accelerator] was constructed Illinois had not experienced any great success in bringing AEC research projects to the state." The editorial went on to chide Illinois Senators Everett Dirksen and Paul Douglas for their membership in the Midwest congressional alliance that was supporting MURA. "Instead of burying the hopes of Illinois for atomic research," it concluded, "they should work to forward the state." Thus were the citizens of Chicago tutored and advised on the intricacies of high energy physics.[9]

Argonne seethed and MURA hoped. At the same time, the midwestern congressional delegation continued to lay rhetorical barrages on the Kennedy administration, and the federal budget for fiscal 1965 inexorably moved through the channels of government, en route to the Government Printing Office. The high energy community, though scattered among a score of research centers in the United States and abroad, was highly inbred, and possessed a marvelous network for conveying any wisp of information. These scientists were among the most affluent of the postwar scientific community, forever on the go— courtesy of government subsidy—to meet each other at conferences around the world; and when they stayed home, transcontinental or transatlantic telephone chats were no novelty. Rumors skipped from AEC headquarters, in suburban Washington, to the Lawrence Radiation Laboratory atop the hill overlooking the Berkeley campus of the University of California; from CERN, the European high energy research center near Geneva, to Dubna, near Moscow, which had become a regular port of call for the nomadic high energy specialists of the west.

As autumn began, it was rumored that Kennedy had privately assured Hubert Humphrey, leader of the protesting midwestern congressmen, that MURA would indeed be given a place in the new budget. Humphrey offered no confirmation of these reports, but, then, the protocol of budget-making accorded the White House first option on news-making announcements. It might choose to share these with favored legislators, or even permit them to reap all the publicity, but, at that point in the budget cycle, the absence of firm news held no significance. In late October, Kennedy spoke at the centennial celebration of the National Academy of Sciences, and thus revived memories of the manner in which Eisenhower had used the occasion of a scientific assemblage to announce the approval of SLAC. But Kennedy merely reaffirmed what the basic scientific community already knew: his administration was aware of the troubled political childhood of basic research in the United States; and, while he was interested in employing science for utilitarian purposes, he accepted the view that basic research thrived best, and paid off in most utility, when left alone. Citing the prewar poverty of basic science, he declared, "If I were to name a single thing which points up the difference this century had made in the American attitude toward science, it would certainly be the wholehearted understanding today of the importance of pure science. We realize now that progress in technology depends on progress in theory; that the most abstract investigations can lead to the most concrete results, and that the vitality of a scientific community springs from its passion to answer science's most fundamental questions." [10] But of MURA, he had nothing to say.

And the reason for this was simply that, within the administration's uppermost scientific and budgetary echelons, sentiments had not yet crystallized into a clear-cut verdict. The Bureau of the Budget, which served as the President's agent for promoting purpose, coherence, and economy in federal spending, argued for canceling MURA, on the ground that the Ramsey panel had accorded the machine a relatively low priority. The AEC, which would be paying the huge costs if the machine received approval, was riddled with conflicting sentiments. On the one hand, its decade-long support of the MURA designers carried an implied commitment of construction funds upon completion of a scientifically acceptable design. That was the way things had worked with all the other accelerators for which the AEC had supplied design money. On the other hand, as a consequence of policies instituted by the Kennedy administration, the AEC was now entering into a new, peculiar, and especially difficult phase in its relationship with the powerful Joint Committee on Atomic Energy. These policies involved cost-effectiveness techniques and also what came to be referred to as "the requirements merry-go-round." In assessing the burgeoning of federal support for research and development during the 1950's and early 1960's, Kennedy's science advisers noted, first, that the bulk of the growth had gone for development, and, second, that enormous sums had been expended on developing hardware which served no clearly defined need or which offered little or no advantage over existing equipment. Prominent among these were virtually identical rocket development programs carried on by the Air Force and the Army and the ill-fated effort to build an atomic-powered aircraft.* From these

* Air Force Secretary Harold Brown, a nuclear physicist who served for five years as director of Research and Engineering for the Department of Defense, stated the issue well, when he wrote, "The development community argues too often for going ahead 'because you can do it.' This is a fine reason for mountain climbing, but not for multi-billion dollar development programs." ("Planning Our Military Forces," *Foreign Affairs*, January 1967, p. 285.)

experiences grew the dictum that costly development programs would not be undertaken unless there was a need for what they were intended to produce and they were the best means of producing it. This policy, which the Bureau of the Budget began to enforce with considerable vigor in 1963, struck directly at several AEC projects that were fondly regarded by the Joint Committee. The Joint Committee, ever eager to find new uses for atomic energy, had successfully pushed the AEC into programs for developing nuclear-powered rockets and nuclear generators for providing electric power aboard space satellites. These were glamorous and highly publicized ventures, but under the criteria of preestablished requirements and cost-effectiveness, they came out poorly. At a huge cost, contended the administration's science advisers, the nuclear space projects would provide no more than a slight improvement over existing systems. The Committee and its allies responded, in turn, that the utility and value of these systems could not be accurately assessed until they were put into at least experimental operation. The "requirements merry-go-round," it was argued, failed to recognize that the existence of a capability is a spur to inventiveness for applying that capability. The administration replied that the AEC's nuclear space systems were inordinately expensive solutions looking for problems.

Accordingly, while the Joint Committee's watchful presence made it perilous for the administration to apply the knife, these favored developmental projects were carefully placed on a diet of progressive financial malnutrition—at the very same time that the administration's science advisers were scheduling great budgetary increases for high energy physics. And the Joint Committee, as its past history surely indicated it would, reacted with alarm. As the committee chairman, Representative Chet Holifield, of California, later summarized the situation, "the overall budget of the AEC stays about the same, but the high-energy physics part of it is increasing. . . . It is increasing at the cost of applied research and developmental projects, many of them right near the point of fruition." [11] The Joint Committee was not unsympathetic to basic research; rather, it was disturbed by the effects that a rapidly growing high energy budget was having on a virtually static AEC budget. Thus, when the AEC came to the high councils to discuss the fate of the proposed MURA machine, it felt, on one side, the pressure and anguish of the MURA scientists and engineers whom it had courted for a decade, and, on the other, the Joint Committee, which was violently protesting the growth of high energy physics at the apparent expense of its pet development projects. Being of a practical disposition, the AEC, though undecided, was not unsympathetic to the view that the time had come to bring MURA to an end.

As the time for decision drew closer, Wiesner, the most influential and politically sophisticated figure ever to serve as presidential science adviser, correctly saw that the MURA issue would have to be decided on—in the best sense of the term—political grounds. The scientists and engineers had told the politicians all that their professional competence enabled them to tell about the MURA machine. Should the United States government commit itself to an expenditure of $170 million for building the machine? The answer of the Ramsey panel boiled down to, Yes—but not if the costs of FFAG would impede progress to higher energies. The question of whether the FFAG expenditure would have that effect was beyond the vision of science. MURA was now at a point where scientific judgment would have to yield to political judgment. Wiesner advised Kennedy that all the scientific returns were in. Looming in the background was their earlier informal understanding that the mood of politics and the advance of science could be reconciled by building one major accelerator approximately every five years. Now it was up to Kennedy to decide whether the political side of that assessment might be revised. The scientists themselves had accorded top priority to the 200-bev. machine under design at Berkeley. Cost estimates on that gigantic accelerator ran to at least $240 million. Wiesner stood ready to provide and explain all the scientific counsel available—in fact, in mid-November, the Ramsey panel was reconvened for two days to see whether it might restate its position in less equivocal terms. It stated, "The Panel wishes to reaffirm its general position in regard to the value of the high-energy and high-intensity frontiers in elementary particle physics"—and the members went off muttering that the White House would have to make its own political decisions.

Wiesner, in his final recommendations, just prior to Kennedy's death, in effect repeated the position of the Ramsey panel: MURA offered a worthwhile machine, but it was not worth a delay in getting on with the 200-bev. machine. When Kennedy died, the decision had still not been made.

Prior to the assassination, Wiesner had announced his intention of leaving the White House post at the end of the year to become dean of science at MIT. But, as did virtually all "New Frontiersmen," he yielded to Johnson's plea for continuity by agreeing to remain on for at least a few months. Thus, he remained in the middle of the still-unresolved MURA controversy—and, as events turned out, the Midwest, with no justification, came to regard him as the villain of the affair.

It may be speculated that Kennedy, with his romantic disposition to science and his commitment to enhancing the nation's intellectual resources, probably would have decided to go ahead with MURA. It is likely that the decision would have been preceded by a good deal of agonizing and perhaps a bit of political horse trading with midwestern legislators who were recalcitrant on matters of concern to the President. Speculating still further, it is doubtful that when Johnson came to the presidency he knew very much, if anything at all, about the long, bitter, and complex history that trailed behind a proposed piece of machinery known as the Fixed Field Alternating Gradient Synchrotron. As a Senator and even as Vice President, Johnson had a reputation for looking out for Texas, so it is doubtful that any of midwestern lobbyists for MURA saw any purpose in seeking to enlist his support before he succeeded to the White House. But Johnson was soon to become well tutored in the intricacies of MURA, for scarcely had his predecessor been laid to rest than the midwestern supporters of MURA, now in a condition resembling frenzy, assiduously undertook an even more frenetic barrage of the White House. What spurred them on was their realization that the new President had chosen economy, frugality, solicitude for every penny of the taxpayers' money, as the leitmotiv of his administration. For Lyndon Johnson, working from blueprints for an unprecedented national consensus, saw that concern over the growth of federal spending was a common denominator among widely divergent political and economic segments of the national electorate. The symbol of this concern was the approaching $100-billion-a-year mark in federal spending. Regardless of its relevance to the economic health of the nation, this emotion-stirring sum had never been touched in peacetime, and Johnson, the consensus builder, was determined to stay below it—at least in the first year of his presidency—even at the price of turning off lights and federal projects.

In December, with the entire federal budget reopened for a last-minute pruning, MURA was the subject of two lengthy meetings between the President and his science and budget advisers. Fundamentally, there was nothing new to be said about MURA, but for the new President it had to be said all over again. Budget Director Gordon and his deputy, Elmer B. Staats, quoted the Ramsey report in reaffirming their opposition to MURA. Wiesner restated the long-standing position that derived from the Ramsey report: the FFAG was desirable, but not at the cost of delaying progress to higher energies. AEC Chairman Glenn T. Seaborg took the same position, but appended to it the view that if the FFAG were built, costs and management problems might be reduced by locating the machine at Argonne. Johnson was reminded that Senator Proxmire of Wisconsin, in his capacity as a member of the Senate Appropriations Committee, had written to Wiesner in September, strongly suggesting that failure to approve funds for MURA "would seriously compromise the prospects" for the higher energy machines favored by the Ramsey panel. Having hinted at political blackmail, Proxmire sought to soften the matter by stating, "I say this not with any notion that there might be some kind of political retaliation. I say this from the standpoint of realism." [12] Johnson's reaction, upon being informed of this crude threat, is not known. But, as Majority Leader of the Senate, Johnson had no affection for the often-erratic, bombastic, and rarely influential Senator from Wisconsin, and to the extent that Proxmire's support for MURA was noted by the President, it is doubtful that it weighed in MURA's favor. However, whatever the factor played by personal antipathies, which, most likely were canceled out by Johnson's long-standing and generally harmonious ties with Humphrey, the fact of the matter was that in the atmosphere of frugality that Johnson was creating, the misfortune-ridden MURA project stood little or no chance of surviving. However, a bizarre series of events and a major political explosion were to accompany its demise. And, as it turned out, though the midwesterners were to lose an accelerator, they were to gain a political principle of infinitely more value than a single accelerator.

Following two lengthy briefings on the complex history of the national high energy physics program, Johnson instructed Wiesner to bring a delegation of MURA representatives to the White House for discussions on Thursday, December 19. At the same time, he directed Wiesner, who had earlier prepared a memorandum stating the pros and cons of the MURA proposal, to prepare a memorandum stating only the arguments against MURA. When Wiesner protested that his own position was one of conditional support for MURA, Johnson is reported to have reminded him that he, Wiesner, had written "a pretty good memo about why we shouldn't build Rover"—a nuclear powered rocket, dear to the Joint Committee—which was being attacked on the grounds of poor cost-effectiveness and limited requirements. Wiesner immediately informed Seaborg of the President's desire to meet with a MURA delegation. Seaborg, in turn, called MURA President Elvis Stahr, who later related, "I was called Tuesday night by Glenn Seaborg of the AEC and told that he and Jerry Wiesner had just mentioned it [MURA] to the President, that they were for it, but that the President wanted to hold down his budget, and wanted or was willing to see one or two spokesmen for 'the proponents'—no politicians should be included." [13]

Thus, the MURA controversy moved toward a climax. Wiesner prepared the memorandum requested by the President and Stahr assembled a delegation, consisting of himself, Bernard Waldman, a physicist on leave from Notre Dame who served as MURA's full-time staff director, and Edwin L. Goldwasser, a physicist at the University of Illinois. However, as events worked out in the frantic schedule that prevailed in the White House less than one month after the assassination, the President was unable to meet with the group as scheduled and the meeting was put off to the next day, Friday.

Now, whether by design or coincidence, Johnson was also scheduled to meet Friday with a small delegation of midwestern legislators, headed by Senator Hubert H. Humphrey, who wished to discuss MURA with the President. That meeting, as it stood on the presidential calendar, was to take place immediately after the re-scheduled session with the MURA representatives. In any case, when Wiesner, carrying the memorandum that he had prepared at Johnson's request, and accompanied by Seaborg, arrived in the anteroom of the President's office, he not only encountered Stahr, Waldman, and Goldwasser, but also a group that included Humphrey, Senator William Proxmire and Representative Robert Kastenmeier, both of Wisconsin, and Representative Melvin Price, of Illinois.

The account of what ensued must necessarily be anonymous, but there is no doubt as to the full accuracy of the events it describes or the approximate accuracy of the conversations quoted:

Leaving both groups in the anteroom, Wiesner went into the President's office to deliver the anti-MURA memorandum (to which he had appended a note restating his qualified support for the midwest machine).

The President, informed by a staff aide that the MURA group was to be followed by the midwestern congressional delegation, was heard to say, "What the hell did you invite those congressmen up here for?"

Wiesner replied, "I didn't invite them. I didn't know anything about it."

To which Johnson said, "Well, goddammit, somebody invited them up." Johnson, burdened by matters more critical than MURA, told Wiesner that he had planned to spend no more than twenty minutes with the MURA representatives. "All right," Wiesner was heard to say, "I'll tell the congressmen that you can't see them." Johnson, the veteran of the legislative process, immediately replied that the waiting congressmen must not be dispatched after having received an appointment to see the President. "You can't do that," he is reported to have told Wiesner. "Bring them in all at once." Thus there assembled in the President's office Wiesner, Seaborg, Stahr, Waldman, Goldwasser, Humphrey, Proxmire, Kastenmeier, Price, and several others.

Johnson started the conference by describing himself as "the only man in government who wants to save money and here are all these people who want to spend money."

Referring to debates within the administration over cutting down plutonium production at the AEC's huge installation in Hanford, Washington, Johnson pointed to the AEC chairman and said, "There's Glenn Seaborg. He wants to run a nuclear bomb WPA out in the State of Washington to get [Senator Henry] 'Scoop' Jackson reelected. . . . Seaborg wants to buy votes. Nobody ever had to buy projects down in Texas to elect me. Not only that, he wants to build a nuclear rocket, and he knows not who and he knows not where."

Johnson went on to say that he was "willing to support anything that science needs," but that the American people expected him to avoid unnecessary expenditures, and he was not persuaded that the MURA machine was needed.

Stahr protested that the machine was indeed needed, and that, while the ultimate construction cost was estimated at $170 million, all that was required in the fiscal 1965 budget was a mere $3.5 million for advanced design and architectural studies.* Humphrey followed this with the exclamation, "Why, my God, the Midwest has been getting shortchanged." Wiesner responded that he had looked into the distribution of funds for academic research and had found no justification for this charge. Humphrey and Wiesner proceeded to talk at each other, when the President interrupted with, "The twenty minutes are up. I've got a memo here that I'd like to read and I'd like you to tell me what you think of it." And he read the anti-MURA memorandum that Wiesner had prepared at his direction. Total silence fell upon the group as the President recited a prosecuter's case against MURA. At the conclusion, Johnson said, "What have you got to say about that?" Those who were present recall that a babble of responses broke out among the dozen or so persons in the President's office. At this point, Johnson abruptly left the room, and the two delegations were ushered out. MURA, born ten years before in a garage near the University of Wisconsin, had appealed for survival to the President of the United States, and had failed.

It is doubtful that Lyndon Johnson, less than one month in the presidency, was fully attuned to the political and emotional implications of the verdict that he apparently intended to issue on MURA, for he was soon to face the realization that, rather than having disposed of the issue, he had actually inflamed it to the point where it might easily interfere with his consensus designs.

On the day following the abruptly terminated White House meeting, Stahr dispatched an angry call to arms to Humphrey, other members of the midwestern legislative delegation, and the presidents of all MURA universities. "I was shocked and stunned by the turn of events just before the sudden conclusion of our meeting at the White House Friday morning," he wrote Humphrey. "The President seemed at times to feel that we were all talking about some sort of toy which it would be very nice for us to have but which the Midwest really doesn't need. . . . But—it *isn't* just 'something new that's nice to have'—it's something scores of scientists have worked on for over several years,

---

* In fields outside of research and development, this argument might have been persuasive. But Johnson, who had long been associated with the space program, knew that cost 'estimates in R and D usually turned out to be far on the low side, and that a good portion of annual R and D expenditures were devoted to projects that had commenced with modest sums.

and which is vital to us; and, if all factors (not just the 'scientific' factor as 'assessed' in the memo) are considered, it's a high priority for the President and the nation *also*. I'm so sure of this, and so fond of him, that I hope you'll find a way to tell him so, soon. . . . It's not too late to get a line inserted in the budget during galley-proofing."

To demonstrate the bipartisanship of support for MURA, Stahr dispatched his most anguished letter to Representative Charles Halleck, of Indiana, Republican Minority Leader of the House. Opening with "Dear Charlie," and stating, "I'm about as angry as I've ever been," Stahr declared, "I'm almost sure *Wiesner* wrote that memo on which the President appeared to be relying. The President knows nothing of the implications of that—or the long history of infighting by East and West scientists to stall MURA. . . . Wiesner, returning as he is to MIT in a month, should have disqualified himself *at least*—and *certainly* failing that, should have passed to the President objectively the scientific advice of the President's Scientific Advisory Council [sic; presumably a reference to the Ramsey panel's qualified endorsement of the FFAG] in up-to-date form, instead of merely quoting the somewhat weasel-worded (though still *affirmative*) recommendation of last April and then trying to show with half-truths why it shouldn't be done. . . . This is not a party matter," Stahr told the Republican minority leader, "it's a Midwest matter, regardless of party. . . . If you can get the President to reconsider—I assure you it's important enough long range to be worth a *major* effort on your part. Also, is there anything else *I* can do for you?" [14]

Since MIT was remote from both MURA and the Berkeley machine that had been given priority over MURA, it is difficult to see why Wiesner's imminent return to MIT should have been grounds for disqualification as presidential adviser on the controversy. But the Midwest, with its *idée fixe* of maltreatment in the disposition of federal research funds, and its obsession over MURA, tended to see conspiracy as the cause of any event that did not promote federal funds for MURA. In their state of agitation, the midwesterners easily forgot that three midwesterners, sitting on the Ramsey panel, had concurred with the preference for high energies.

Two days after his impassioned letter to Halleck, Stahr wired the MURA university presidents that following telephone conversations "with Humphrey and others it appears that we may still have a chance of getting a favorable decision if strong views as to the importance of the projects to the future of midwestern science, and, therefore, midwestern higher education can be registered with the White House by midwestern educators and Congressional delegations during the next several days. . . . Please ask your senators and congressmen to send wires urging the President not to strike a blow against the midwestern economy by killing the MURA project. We are assured that all telegrams will be brought to the President's attention. . . ."

In late December 1963, when the last type was set on the budget for fiscal 1965, MURA was beyond resurrection. Johnson, in the first days of his presidency, could not permit political pressure to detract from the image of frugality that he was carefully projecting. In addition, the administration's scientific and budgetary advisers no longer had energy or patience for the seemingly interminable and enervating struggles generated by the midwestern design group. The President had decreed an end to MURA, and no one around him felt any inspiration for proposing that the issue be reviewed still once again.

On the other hand, the political resentments engendered by the MURA decision were not to be ignored—and they were not ignored. On January 16, 1964, after the midwesterners had been formally advised that AEC support for MURA would be terminated with a $500,000 allotment to cover the cost of going out of business, Johnson wrote to Humphrey, "I fully understand your disappointment with the outcome. The decision was a difficult one. I devoted more personal time to this problem than to any nondefense question that came up during the budget process." But, he added, "Given the scope and scale of the investment we have made and will be making at Argonne, I found it impossible to justify starting another national laboratory close by. . . . I share fully your strong desire to support the development of centers of scientific strength in the Midwest, and I feel certain that with the right cooperation between the Government and the universities we can do a great deal to build at Argonne the nucleus of one of the finest research centers in the world." [15]

Four days later, the AEC issued a press release worthy of the State Department's most obfuscatory diplomatic communique writers. Titled "Plans for Continued Support of Accelerator Design Announced," it disclosed that the MURA group would be given an opportunity to relocate at Argonne to work on the 600–1000-bev. machine that was in the early design stages at Brookhaven. Few of the MURA scientists took up this offer, and, in fact, MURA effectively disintegrated shortly after the rejection of the FFAG design. Most significantly, however, the announcement stated that, while design work was proceeding at Brookhaven and at Berkeley, where the 200-bev. machine that had received the Ramsey panel's preference was nearly ready for construction, "No decisions have been made for the construction of either of these large national accelerators, nor have site locations been selected." Finally, in the seventh and final paragraph of the press announcement, the AEC provided MURA's obituary notice: "The reorientation," it stated, "follows a decision not to construct a 10–12.5-bev. high intensity accelerator which has been proposed by the MURA group." [16]

The President had partially explained his decision against MURA on the grounds that he could not justify another national laboratory near Argonne. Two years later, however, after strenuously combing sites throughout the nation, the AEC decided that the proper site for the 200-bev. machine, designed by Berkeley's Lawrence Radiation Laboratory, was Weston, Illinois, twenty-five miles from the Argonne National Laboratory.

The legacy of MURA was a new politics of science.

NOTES

1. *Equitable Distribution of R and D Funds by Government Agencies, Hearings,* Subcommittee on Government Research of the Senate Committee on Government Operations, July 25–27, 1966, p. 20.
2. *Federal Support for Academic Science and Other Educational Activities in Universities and Colleges, Fiscal Year, 1965,* NSF Publication 66-30, p. 21.
3. *Congressional Record,* July 2, 1962.
4. "The Changing Patterns of Defense Procurement," mimeographed report, Department of Defense, June 19, 1962, p. 8.
5. *Charter of Progress* (Dallas: Graduate Research Center of the Southwest).
6. The Midwest Resources Association, proposal for the location of the Atomic Energy Commission 200-bev. accelerator, 1966.
7. Charles Kimball, president, Midwest Research Institute, remarks to Midwest Governor's Conference, September 10, 1964.
8. Albert Shapero, "Impact of Federal Support for Science and Technology on Regional Economic Development," testimony before Subcommittee on Government Research of Senate Government Operations Committee, May 11, 1967.
9. Chicago *Sun-Times,* December 22, 1963.
10. John F. Kennedy, "A Century of Scientific Progress," *The Scientific Endeavor* (New York: Rockefeller Institute Press, 1965), p. 311.
11. *AEC Authorizing Legislation,* 1965, p. 1487.

12. Letter, Senator William E. Proxmire to Jerome B. Wiesner, September 27, 1963.
13. Letter, Elvis Stahr to Charles A. Halleck, Republican Minority Leader, December 22, 1967.
14. *Ibid.*
15. Letter, Johnson to Humphrey, January 16, 1964.
16. AEC Press Release, G-14, January 20, 1964.

XII

The New Politics
of Science

Science . . . can no longer hope to exist, among all human enterprises,
through some mystique, without constraints or scrutiny in terms of
national goals, and isolated from the competition for allocation of
resources which are finite.

IVAN BENNETT, Deputy Director,
Office of Science and Technology, 1966

To comprehend the new politics of science, it is necessary to recognize
a fundamental fact about the old politics. And the fact is this: The old
politics, on the whole, had served science very well. Despite the political community's refusal to implement Bush's grand design, despite
upheavals and debacles such as Mohole and MURA, the old politics
had enabled American pure science to move from poverty to affluence,
from weakness to strength. Government never wholly accorded the
sovereignty demanded by the ideology of pure science. But though reins
and restrictions existed, and the principle of accountability (loathsome
to the scientists) was never absent, the essential point was that, *in
practice,* scientists wrote most of the rules for the use of federal research
money; scientists staffed the agencies that dispensed the money, and
scientists from the university community advised these same staff scientists on the distribution of the money. As Don K. Price, dean of the
Faculty of Public Administration of Harvard, observed in an essay
published in 1962, "the plain fact is that science has become the major
Establishment in the American political system: the only set of institutions for which tax funds are appropriated almost on faith, and under
concordats which protect the autonomy, if not the cloistered calm, of
the laboratory." [1]

Government traditionalists and guardians against conflict of interest,
as well as plain skeptics, express outrage at this overlapping of roles
in the relationship between private supplicant and public patron. And,

in terms of the adversary process that is at least implicit in the traditional procedures for dispensing federal funds, there are ample grounds for outrage. Nevertheless, though science's peculiar linkage to government did more than a little violence to democratic processes, it passed one important test: It worked very well, as measured in what was good for science. By almost any indicator—Nobel prizes, quantity and significance of research papers, direction of flow of the so-called brain drain—American science had achieved a splendid condition by the early 1960's.* Though Soviet testimonials to the quality of American science, and American testimonials to the quality of Soviet science, should be scrutinized with care as to motivation, it is noteworthy that in recent years Soviet scientists have openly conceded that the administrative diversity and amorphousness of American science have proved superior—in terms of scientific quality—to their own rigid, centrally administered system for running pure research. Thus, in 1966, Peter Kapitsa, director of the USSR's Institute of Physical Problems, and a member of the Presidium of the Soviet Academy of Sciences, lamented that, though the scientific communities of the two nations were similar in size, United States scientists produced one-third of the reports in the world's major scientific journals, while Soviet scientists produced only one-sixth. Soviet pure science, he argued, was burdened by administrative criteria and procedures that impaired the functioning of research. Research directors, he said, lacked freedom to staff their laboratories on the basis of professional competence; efforts were made to run basic research by edict, rather than by permitting researchers to pursue their own lines of investigation. Kapitsa also complained that "delegates [to international scientific meetings] are selected by bureaucratic methods, without taking into consideration scientific qualifica-

---

* One reasonably good measure of the significance of a scientist's work is in the frequency with which other scientists refer to it in their own work. Using this measure, it is illuminating to note the findings of the Westheimer study of chemistry (see Chapter VIII). In 1962, American scientists produced approximately one-quarter of the chemistry papers and patents listed in *Chemical Abstracts,* which systematically covers the chemical periodical literature in approximately 100 countries. In the major chemistry journals of the British Commonwealth, reference to "domestic" papers totaled 1,048, while references to papers produced by American scientists totaled 1,175. References to all other foreign papers totaled 955. In the Soviet chemistry literature, domestic references exceeded those from the U.S., 583 to 234, but in Japan, references to American work slightly exceeded the domestic total, 162 to 160. (*Chemistry: Opportunities and Needs,* National Academy of Sciences–National Research Council, 1965, p. 32.)

tions and interest." [2] In short, Kapitsa, in harmony with what many of his own Soviet colleagues had been saying for years, was arguing for the flexibility, the formlessness, and the opportunities for scientific entrepreneurship on which American pure science had thrived for years.

The financial basis for American pure science's ascent to excellence was federal money—money that, with trivial exceptions, was either unwanted or unobtainable by the practitioners of pure science prior to World War II. But though science and government had achieved financial intimacy, they had not resolved the ancient tensions arising from government's insistence upon accountability and science's desire for independence, government's preference for utilitarian research and science's addiction to indulging its own curiosity. Rather, over two decades, they had muffled and baffled these tensions with a wondrously complex patchwork of functional fictions and administrative, economic, and political contrivances. If political support could not be mobilized for basic science as a worthy activity in and of itself, then a Cold War motivation would be introduced and exploited. If Congress refused to share the scientists' vision of NSF as the principal fount of government support for academic basic research, then the Defense Department would pay a major portion of the bills—under the guise of research essential to the military. If political interference and pork barrel instincts threatened to interfere with the distribution of research funds on the basis of scientific quality, then the project and peer systems would be deified as the means for distributing money for the support of research. For the purpose of protecting the quality of science—which was a principal concern of the scientific architects of the system—it was vitally important to maintain the project and peer techniques, regardless of their adverse effects on lesser-ranking institutions, on the relationship between teaching and research, and on the internal coherence of universities staffed by scientists whose financial lifelines ran directly to Washington.

Thus, two decades after Bush and his colleagues had brought the penurious scientific community to the wartime service of government, American science had become affluent, highly productive, and the *de facto* sovereign of its own most vital affairs. At the same time, it had also become ripe for reaction from its surrounding environment. That reaction, coming in bits and pieces, has not yet fully spent itself. But in this concluding chapter an attempt will be made to identify its most significant elements, describe their genesis, assess their course and effects, and examine some steps that might, hopefully, accrue to the benefit of both science and society.

By the early 1960's, something of a qualitative change began to manifest itself in the political community's attitudes toward science, technology, and the statesmen who represented them in Washington. To a degree, this change was a repetition of earlier history. World War II had established the romance of science and government. By 1957, the honeymoon was long past, and, though the permanence of the relationship was beyond doubt, the effects of familiarity were taking a toll. Not untypical of this period and its mood, Secretary Charles E. Wilson, presiding over that mainstay of financial support for basic research, the Defense Department, proclaimed, "Basic research is when you don't know what you're doing." [3] Taking a lead in one of the Eisenhower's recurrent, though ineffective, economy drives, Wilson announced a cut of 10 percent in Defense support for basic research, though it was estimated that, because of lagging budgets and inflation, the Department, in 1957, was actually supporting 25 percent less basic research than it had in 1952.[4] The significance of these estimates is open to question, since, during this same period, other federal agencies, such as the AEC, NSF, and NIH, had sizably increased their spending for basic research. In any case, the Department's scientific clients were energized to a state of panic by Wilson's plans. The Air Force Office of Scientific Research, the struggling counterpart of the Navy's very successful Office of Naval Research, was on the verge of dispatching telegrams canceling 600 research projects, most of them basic research projects in universities, when it tried a last desperation move. It appealed to Rabi, Wiesner, and DuBridge. Rabi, who had succeeded DuBridge as chairman of what was then the executive branch's highest ranking science advisory body, the Science Advisory Committee of the Office of Defense Mobilization, took the case to the White House, and returned with reports of a sympathetic response.[5] But the appeal was almost immediately rendered unnecessary. In October 1957, the Soviets orbited Sputnik, and honeymoon fervor immediately returned to the romance between science and government.

Fear of Soviet superiority in science and technology, regardless of what the realities might be outside of the USSR's clear-cut superiority in space, instantly overwhelmed the spirit of economy, and opened the way for a new phase in federal relations with the intellectual realm. Thus, in 1958, the passage of the cleverly titled National Defense Education Act provided extensive federal aid to education not directly related to the established favorites, science and technology. But the political impact of Sputnik was most strongly felt in these two areas, and resulted not only in a vast increase of federal financial support for research and scientific training,* but also in the re-enthronement of scientists in the positions of public acclaim and government influence that they had first occupied at the end of World War II. From 1945 on, while their presence had increased in Washington, so had the familiarity with which they were viewed. By 1957, scientists in public affairs were no longer a novelty, and were no longer accorded the reverence that had been bestowed upon the bomb and radar builders when they emerged from their secret wartime laboratories. Rather, a dozen years after the end of the war, scientists had become familiar parts of the Washington landscape, and, as Wilson's economy drive suggests, their profession began to suffer the burdens of being viewed as just another supplicant for federal largesse. Sputnik, in a moment, sent the pendulum swinging back. Not only did it inspire the release of a huge flow of funds, but it also inspired government to institutionalize many parts of the science advisory system that had randomly sprung up throughout Washington after the war. The Science Advisory Committee, buried within the obscure office of Defense Mobilization, was elevated to the status of the President's Science Advisory Committee (PSAC). Killian, long a part-time adviser, was called down from MIT to head the committee on a full-time basis, and also to serve as the President's special assistant for science and technology.† Thus, cour-

---

* The Air Force Office of Scientific Research, for example, had been advised, pre-Sputnik, that its budget would be severely cut. Almost immediately after Sputnik went aloft, OSR's budget was raised 38 percent over the previous year's figure. (Nick A. Komons, *Science and the Air Force, A History of the Air Force Office of Scientific Research,* Historical Division, Office of Aerospace Research [Arlington, Va., 1966], pp. 108–9.)

† It is not unlikely that Rabi, as chairman of the Science Advisory Committee, could have become, if he so chose, chairman of the newly established PSAC. But Rabi was the epitome of the commuting academic statesman of science. Though deeply involved in Washington affairs, he never held full-time government employment, except for his service at the MIT Radiation Laboratory during World War II. His home base was the Columbia University physics department, and though his government advisory activities often took him to Washington several days a week, year in year out, he always returned to his academic setting.

tesy of the Soviet space program, the science-government relationship resumed honeymoon status. But Sputnik, though a powerful aphrodisiac, was not a permanent one, and by the early 1960's, reaction once again set in.* This time, however, it did not simply manifest itself in financial ways, though money troubles were certainly part of the picture. Rather, by the early 1960's, when the Mohole and MURA projects, both nourished by post-Sputnik affluence, were moving toward their unhappy and well-publicized conclusions, the political community's interest in science began to penetrate to previously untouched areas within the scientific community. In its earlier dealings with basic research, Congress had generally confined itself to determining budgetary totals, usually employing no more than a seat-of-the-pants system that provided each year for spending a little more than was spent in the previous year. But, by 1963, when the federal R and D budget had reached $13.7 billion and it was estimated that it would rise to $16.3 billion the following year,[6] something new appeared in the picture. Congress not only stood by its traditional concern for how much money was being sought for science and technology, but it also began to probe into why it was sought, where it would be spent, under what circumstances, and with what assurances of accountability and value returned. Basic research, of course, regularly pleaded that it received no more than 10 percent of total federal expenditures for R and D, and that it should not be penalized for the burgeoning of federal expenditures for military and space hardware. But pure science had never fully educated its congressional masters to the distinction between basic research and technology. Thus, in 1962, NSF Director Waterman protested that not all R and D expenditures were for basic research, "I know," replied Senator Allen Ellender, "but it's all science." [7]

* The fortunes of the science advisory office in the State Department symbolizes the cyclical nature of the science-government relationship. In 1951, upon the recommendation of a study group headed by Lloyd Berkner, a science advisory post was created in the State Department. By 1953, it had failed to take root and was placed on what was essentially a caretaker basis. Immediately after Sputnik, it was deemed unthinkable for the nation's principal agent of foreign policy to be unattended by a statesman of science, and the office was revived. However, in 1964, the physicist who had held the post for two years returned to academe; a nonscientist State Department civil servant eventually took his place on an acting basis—and nearly three years later formally succeeded to the position. In the mid-1960's it had become thinkable for the State Department to be unattended by a ranking scientist.

The main cause of the swollen budgets that aroused Congress was simply modern govenrment's need, or, at least, presumed need, for costly technology. But the old politics of pure science—predicated on government money and scientific sovereignty—could not sustain the scrutiny of a Congress that tended to lump together all science and technology, and hold them to standards of accountability that, though anathema to the ideology of science, were central to government administration. And the principal product of this scrutiny was a new politics of science, a politics characterized by a diminution of the *de facto* sovereignty that pure science had nurtured throughout the postwar period. By the early 1960's, science, viewed as one with technology, had become too expensive, too poorly understood and misunderstood in terms of its intrinsic and economic values, to be left to the scientists. Thus began the new politics of science. The outcroppings of this politics are many and still forthcoming, but, when viewed against the inquiry that we have so far conducted into the postwar politics of pure science, a final case study, this one involving medical research, should suffice to convey its characteristics and significance.

In 1959, the rapid growth of federal support for medical research attracted the attention of one of Congress's "oversight" bodies, the House Intergovernmental Relations Subcommittee. Under the general mandate of its parent group, the Government Operations Committee, the subcommittee, chaired by L. H. Fountain, of North Carolina, held authority to investigate whether government agencies were performing according to legislative intent. Since congressional forced-feeding had quadrupled the annual NIH budget, to a figure of $213 million, between 1950 and 1957, it was a reasonable assumption that, amidst this sudden gusher of revenues, NIH might not be dispensing its wealth in detailed accordance with its own regulations and pertinent government-wide regulations. After two years of poking and probing (during which time the NIH budget rose to $577 million), the Fountain committee issued a mildly worded report recommending that NIH generally tighten up its procedures for assuring that its funds were being used for the purposes for which they were dispensed.[8] In terms of the traditional politics of science, the most significant of these recommendations rejected the peer system as being adequate by itself for governing the distribution of NIH funds. The work of NIH's outside advisory committee—the so-called study sections that assessed the scientific validity and importance of grant applications—"should be complemented" the subcommittee stated, "by a thorough review of each project's financial requirements performed by qualified [full-time government-employed] analysts in the Division of Research Grants." In explaining this recommendation, Fountain's report noted that "the study sections do concern themselves with the reasonableness of budget requests in relation to the work proposed. . . . This, however is not the type of systematic budget examination that is required to satisfy NIH's administrative responsibility." *

In these few words, Fountain was chopping at the ventricles of the economic system that pure science had laboriously assembled in the postwar period, for, in effect, he was contesting the traditional scientific view that the internal value system of science guaranteed an ethical standard that required no outside surveillance or reinforcement.

Following Fountain's 1961 report, NIH and its administrative senior, the Public Health Service, did virtually nothing—as they later acknowledged—beyond lavishing upon Fountain fulsome praise for his interest in their activities. (The Surgeon General, who is head of the PHS, told Fountain, "The intelligent examination of recent practices and the thoughtful recommendations for their improvement are very helpful in focusing attention on problem areas and suggesting the need for revised procedures. I want to assure the committee that each criticism is being most carefully examined and each recommendation most seriously considered both in my office, and the Director of the NIH, and by those immediately responsible for the grant administration at NIH." [9])

But, in fact, nothing happened, for, while Fountain, in 1961, was

* Fountain's report also concluded that "Extravagance and irregularities have been found in the handling of grant funds by conference planning groups," and generously proposed that NIH itself formulate guidelines for travel to domestic and international meetings. There is no doubt that travel funds had been misused to a not insignificant extent, under the guise of attendance at professional meetings. But the plentitude and free use of these funds also assured that scientists could easily attend meetings of importance to the progress of their research. While the Soviets complained (see p. 270) that "bureaucratic methods" affected international travel of their scientists, without "taking into consideration scientific qualifications and interest," the worst that could be said about the American system was that some chiselers could easily worm their way in among qualified scientists who had legitimate reasons for traveling to professional meetings. Contemplate the difficulty of separating the disguised tourists from the legitimate scientists, and take your choice of method.

criticizing the manner in which NIH accounted for the use of its money, his congressional colleagues were pouring funds into NIH, above and beyond even the generous sums specified in the presidential budget. Under these circumstances, and burdened as they were by the task of dispensing funds in response to some 15,000 separate grant applications per year, it is not surprising that NIH administration had little time or motivation to pay attention to the nonscientific carpings of Congressman Fountain.* NIH officials knew better than the congressman ever possibly could that some leakage was occurring in the use of their funds. But they believed, and probably were right, that the leakage was insignificant when measured against the superb biomedical research enterprise that was being built across the nation with NIH funds.

In 1962, one year after the publication of the subcommittee's first report, Fountain summoned the leaders of NIH for an account of what they had done to fulfill his recommendations. The answer, that they had actually done very little, was frankly stated by James A. Shannon, the director of NIH, and Ernest Allen, the associate director for research grants. As they expressed their reservations about the wisdom of adopting Fountain's prescriptions for the administration of biomedical research, they drew deeper into conflict with the guiding hand of Fountain's investigation, Delphis C. Goldberg, a Harvard Ph.D. in political economy and government, who made little effort to conceal his contempt for the scientists' insistence that the workings of science inevitably assured proper use of the taxpayers' money. There ensued a colloquy that is unsurpassed as a concise, bedrock exposition of pure science's ideology of sovereignty and government's ideology of accountability for public funds.

* A major impediment to tighter administration of NIH funds was the generally antiquated and often chaotic state of university business offices. Though NIH dealt directly with its scientist-grantees in dispensing funds for research, research funds were actually turned over to the scientists' institutions, which, in theory, were supposed to make certain that the money was used as intended. Few university business offices were equipped to handle the great volume of funds dispensed by NIH; furthermore, scientists, in the best academic tradition, often viewed administrators as natural enemies, and if the administrators' rulings interfered with their desires, they were fully capable of invoking their prestige against the campus bookkeepers. In a showdown, they were also in a position to pick up their grants and move to another institution, since, under the project system, it was the man and his project, not the institution that was being supported by NIH.

Allen, in opposing the establishment of an internal NIH review of the study section recommendations, stated, "I feel . . . that the scientific community at large has come to know that it is important to them in the review of their proposals [by the standard study section system] that they provide us with a realistic budget. . . ."

Goldberg: "Are you suggesting that the scientists who prepare budget requests for their [own] research proposals are best qualified to determine the minimum requirements of the projects?"

Allen: "I think they are."

Shannon: "Who else would do it better than they?"

Goldberg: "Isn't it a human tendency for people to ask for the funds they would like to have if past experience suggests these are the amounts they might get?"

Shannon: "Dr. Goldberg, we have never operated our business that way. Nor do we believe that most scientific groups in the country have an asking and a selling price for their product, which is research activity. I think we get a realistic appraisal of what they need to do the job. What they ask for is the result of their best judgment." [10]

Upon concluding those hearings, the Fountain subcommittee emphatically stated its dissatisfaction with NIH's response to the recommendations, and, from that point on, NIH, long the beneficiary of congressional largesse and protection, found itself in a new and increasingly hostile political atmosphere. "The committee is dissatisfied," Fountain's report declared, "with the slow progress which NIH has made to strengthen the management of the grant programs for health research. While NIH has acted in several areas in response to the committee's recommendations, relatively little effort has been made to improve the overall management of these important grant programs. In particular, the committee has found no significant improvement in the inadequate fiscal review of project requirements on which it reported last year.

"In the absence of appropriate policies, procedures, and adequate staffing the nongovernmental scientists who served on study sections are, in effect, determining the budgetary needs of research projects." To which the committee added, ". . . freedom for the scientist should not be confused with license or fiscal irresponsibility. One cannot condone waste and extravagance wherever it exists as being either in the public interest or in the interest of science." [11]

Actually, Fountain's *publicly* cited evidence of "waste and extravagance" was relatively trivial, involving, principally, seamy business practices by a commercial research firm that had received $378,600 from NIH between April 1959 and August 1962. NIH's dealings with profit-making organizations were a rarity, and it was doubtful that, as lax as NIH's principal constituents, universities and medical schools, might have been, that they matched the private firm's cupidity or brazen misuse of research monies. But, always lurking around NIH's dealings with Fountain were Fountain's hints that he knew more than he was revealing and NIH's realization that there was at least a bit, possibly a great deal, more to reveal than had come out on the public record.*

The effects of Fountain's investigations were immediate and are still being felt in the biomedical research community. Having operated throughout the postwar period on a policy based on implicit trust of its grantees, NIH took to appending ever more complicated accountability requirements to its grants.

Shannon, in an appearance before NIH's political guardian and benefactor in the House, Representative John E. Fogarty, clearly spelled out the anguish of his predicament. Acknowledging that "additional steps have to be taken to make certain that the grantees spend the funds awarded to them for the purposes for which the grant was made," Shannon added: "Now contained in that simple statement on the one hand is a situation fraught with hazard for the future of American science and, on the other hand, it is a sound administrative statement." [12]

Nevertheless though the rules often were changed from month to month as NIH sought to maneuver between its bitterly complaining scientific clients and its intransigent congressional critic, there was no

---

* The extent to which scientists have stretched, or ignored, the rules governing use of federal research funds is difficult to assess. It is interesting to note, however, that in 1964, the Committee on Science and Public Policy, of the National Academy of Sciences, stated, in a comprehensive review of government-science relations, "We believe that understanding of the federal support of basic research by the project grant/contract system is not sufficiently widespread in the scientific community. Grants and contracts are given as trusts to institutions for a purpose, which is substantially as described by the principal investigator in his proposal. The investigator assumes a major responsibility in accepting federal funds and has an obligation to account for their proper use. Acceptance of a grant commits him to a conscientious effort to achieve its stated purpose; he acquires no other rights to the granted or contracted funds." (*Federal Support of Basic Research in Institutions of Higher Learning*, NAS-NRC, 1964, p. 7.)

doubt that Fountain's concept of accountability was, on the whole, coming to dominate NIH's relationship with the scientific community. In the past, if a grantee believed that the project for which he had been given funds was not panning out, he was virtually free to take up another. Under the new regulations prior approval by NIH was required. In the past, grantees would simply list a lump sum for "equipment" in their grant applications. Now they were required to specify their equipment needs, and, were prohibited, without prior approval, from purchasing any unlisted item that cost over $1000. Research funds could no longer be used for travel unless specifically authorized in the award of the grant. Travel abroad was held to particularly close account. Finally, since many universities looked upon federal research grants as a convenient source of money for supplementing faculty salaries, it was specified that salaries charged to research must reflect time spent on research, and that the rate of pay for research time must not exceed what the university was paying the grantee for his teaching time.[13] With this regulation, there inevitably followed a requirement that NIH grantees must report periodically, usually quarterly, the amount of time actually devoted to NIH-supported research, though many protested, and with a good deal of justification, that research defies measurement by the clock.* Though NIH never fully responded to Fountain's demand that government employees should effectively be the ultimate authority over the amount of money required to carry out a specific research project, NIH did adopt the practice of generally trimming a good deal from the amounts sought by its grantees.

By nonscientific standards, none of these requirements was onerous. And it would be difficult, perhaps impossible, to demonstrate that they affected the conduct of research beyond the small amount of additional time that they exacted for paperwork. Nevertheless, the biomedical community howled. Typical of the reaction was a letter that Barry

---

* The classic assertion of the scientists' antipathy to timekeeping was produced by an internationally renowned young mathematician, Stephen Smale. When NSF demanded to know how he had spent his time during a period when he was receiving its support, Smale replied that he had attended a professional conference and had also been working on mathematics while holding a visiting appointment at a European university. To which he added, "I was also doing mathematics, e.g., in campgrounds, hotel rooms, or on a steamship. On the S.S. *France,* for example, I discussed problems with top mathematicians and worked on mathematics in the lounge of the boat. (My best known work [for which he received the Fields award, mathematics equivalent of the Nobel Prize] was done on the beaches of Rio de Janeiro, 1960!)" (*Science,* October 7, 1966, p. 132.)

Commoner, a botanist at Washington University, St. Louis, wrote to *Science:* "If, as I believe they will, the new rules force investigators to adopt a more arbitrarily restricted style of work, many present and potential scientists will find a life in science significantly less attractive and productive than it once was. This will damage the future of science in this country by an amount far greater than that involved in the extra paperwork imposed by new administrative requirements." [14] *

However, the effects of Fountain's venture into the administration of biomedical research were soon to extend far beyond the paperwork issue. For the congressman's major accomplishment, intended or not, was to add persuasive evidence to the belief that Congress had been giving NIH too much money too fast. Fountain himself drew upon and reinforced this belief when he introduced into his hearings a Washington *Star* editorial that critically observed: "There probably is no parallel in Government to the grim determination with which Congress forces money on the National Institutes of Health." [15]

The presence of "grim determination" is open to question, but there was no doubt that the political wizardry of Representative John Fogarty and his Senate counterpart, Lister Hill, had created an atmosphere in which opposition to NIH's budgetary desires was equated with indifference to disease. Fountain, however, had deftly skirted the issue of budgets and disease, and, instead, had struck NIH in its most vulnerable area: the system of administration that had evolved under the influence of the ideology of pure science. In effect, Fountain had formulated a respectable means for supporting NIH while opposing its methods of doing business.

Once the meaning of this settled into the U. S. Congress, the effects swiftly spilled beyond the details of administration and NIH's history of fabulous financial growth came to an end. Between 1957 and 1963, the NIH budget had risen from $213 million to $930 million. In fiscal 1964, the first post-Fountain budget, it rose only to $974 million.

---

* The scientific community was, in general, strongly opposed to NIH's new regulations, but not all scientists viewed them with concern. Following publication of a mass letter of protest regarding the regulations, a scientist who served on various research grant committees advised the readers of *Science* that one of the signatories "is a gentleman more than adept at securing grants and then not using them for the purpose for which they were secured. This, of course, is often justified on the basis of 'research,' 'education,' and 'serendipity.' The cold fact remains that power corrupts and large amounts of funds spell power. Eternal vigilance remains the price of liberty and, to say the least, honesty." (L. H. Garland, letter to editor, *Science,* June 7, 1963.)

In 1965, it edged over the $1 billion mark for the first time. By pre-war and early postwar standards, these were stupendous sums. But, rather than joy, a mood of crisis enshrouded the national biomedical enterprise. Confronted by a cooling of congressional support, the politicians of medical research, lay and professional, had nowhere to turn for assistance. For at the very beginning of the postwar alliance between science and government, medical research had gone its own political way. It had sought out and developed alliances with Congress, in contrast to the route chosen by the bomb and radar builders, who had found easy entree to the White House because of their professional skills and wartime experience in strategic affairs. In fact, the physical scientists who dominated the White House science councils had often looked with envy, and sometimes annoyance, at NIH's unique ability to negotiate directly with Congress on budgetary matters. Kistiakowsky, for example, recalls that while serving as Eisenhower's science adviser, he had "virtually nothing to do" with the health-related sciences. "NIH was too powerful," he explained, with admiration. "Shannon had a perfect understanding with Lister Hill and John Fogarty."

The sense of crisis that prevailed when NIH was at the $1 billion level did not derive from ingratitude. Rather, the abrupt leveling off of the financial growth rate came at a time when biomedical research was especially vulnerable to disruptions from downward financial changes.* Buoyed by a uniquely harmonious and profitable relation with Congress, the leaders of NIH assumed that the growth rates of the late 1950's and early 1960's would continue and that, as a result, national spending—public and private—for health-related research would rise from approximately $1.4 billion in 1963 to $3 billion by 1970.[16] In preparation for this anticipated boom, they put great emphasis on programs to train a new generation of scientists. Thus, for 1963, NIH scheduled the expenditure of $190 million for training and fellowships, and additional funds for training were built into its research grants to provide support for graduate research assistants.[17] By the early 1960's, these programs were annually turning out thou-

---

* It is worth emphasizing that the NIH budget did not decline. Actually, the budget went up, but the percentage of the increase was relatively small in comparison to Congress's traditional handling of NIH money matters. In the past, NIH budgets had been increased annually by 15 to 30 percent; following Fountain's investigation, the increase dropped below 10 percent.

sands of well-trained young scientists, and many of them, having been inculcated with the mystique of pure research, were eager to get on with research programs of their own. However, at the very same time, though the money flow for biomedical research was not receding, it was not increasing at a sufficiently rapid pace to accommodate both the existing research establishment, which regularly contended that it needed more simply to stand still, and the newcomers to the ranks of research. On top of this, NIH, along with all other federal research agencies, began to feel the political backwash of the Mohole and MURA controversies. In particular, congressmen demanded to know what NIH was doing to spread its wealth. Shannon frankly replied that, despite the appearance of wealth, NIH did not have the money to support all of the qualified young scientists that had been trained at NIH expense, build up new centers of excellence, and still finance research in the established centers of science.

The old politics of science had guided the construction of a huge, costly, and growth-oriented research establishment. By most standards, it was eminently productive. But, having been nurtured on self-government and uninterrupted growth, it was not prepared for Congress's intrusions into its affairs, the ensuing deceleration of financial support, or the new politics of science which these events symbolized. Nor was any field of research easily to accommodate new difficulties that followed. Along with all other domestic programs, science found itself required to stifle costly ambitions because of the financial burden of the Vietnam War. But, more important, science found itself confronted by a resurrected ancient foe: political preference for utility in research. Lyndon Johnson, dedicated to rapid social engineering in a nation plagued by many ills, chose to reopen the question of the utility of basic research. And, having spent the very first few days of his presidency dealing with MURA and its violent complaints of maldistribution of federal research funds, the President from Texas, alumnus of Southwest State Teachers College, also chose to reexamine the alignment of power within the scientific community.

Within two years after Johnson's succession to the presidency, a number of little-noted, but highly significant, changes had occurred in the relationship between the scientific community and the United States government. Since its creation in 1957 the eighteen-member President's

Science Advisory Committee had contained at least half a dozen members from the Harvard and MIT faculties. However, under Johnson, as the terms of the Cambridgeites expired, their replacements were drawn from other parts of the nation—not from any particular section, but simply from regions beyond Cambridge—until, by 1966, the Harvard-MIT representation was reduced to one. In the summer of that year, Senator Fred Harris, of Oklahoma, chairman of the Government Research Subcommittee, one of several newly created congressional committees that had sprung up as part of the new politics of science, held an off-the-record meeting with some dozen scientists and engineers to discuss federal research programs. One of the participants, a PSAC alumnus who was perhaps the most influential of the Harvard-MIT science advisory veterans, wisecracked at the outset, "There has been no presidential directive as to the number of speakers you can have from Cambridge, has there?" [18] No response was made or needed. The Cambridgeite had made an "in" remark to an "in" group. All were aware of the presence of Lyndon Johnson in the politics of science.

The old politics of science was an elitist politics. For the financing of science, the old politics had created a self-reinforcing system of quality, whereby those institutions that had achieved excellence were given the sustenance to achieve still greater excellence. Leftovers were provided for secondary institutions, but whenever quality sprouted on their premises, it was always vulnerable to raids by the heavily subsidized clients of the system. The old politics of science was also an introverted politics. It argued, and sincerely so, that what was good for science was good for society; that when science thrived—and the contention, of course, was that it thrived best when left alone—society inevitably benefited.

Elitist, introverted—and successful—the old politics was also untenable in the presidency of Lyndon Johnson, a people's President. Six weeks short of the completion of his second year in office, Johnson issued a declaration that, in effect, certified the demise of the politics that had governed the science-government relationship since World War II. It is worth quoting at some length, for between its lines were strewn the bones of the old politics of science:

"Our policies and attitudes in regard to science cannot satisfactorily be related solely to achievement of goals and ends we set for our research. . . . We must, I believe, devote ourselves purposefully to developing and diffusing—throughout the nation—a strong and solid scientific capability, especially in our many centers of advanced learning. . . . At present, one-half of the Federal expenditures for research go to 20 major institutions, most of which were strong before the advent of Federal research funds. During the period of increasing Federal support since World War II, the number of institutions carrying out research and providing advanced education has grown impressively. Strong centers have developed in areas which were previously not well served. It is a particular purpose of this policy to accelerate this beneficial trend since the funds are still concentrated in too few institutions in too few areas of the nation.

"Under this policy, more support will be provided under terms which give the university and the investigator wider scope for inquiry, as contrasted with highly specific, narrowly defined projects." [19]

Amidst the conditions that produced this proclamation, and that, in turn, were reinforced by it, the new politics of science have emerged into full bloom. Thus, when the Lawrence Radiation Laboratory, at Berkeley, completed designs for the 200-bev.-accelerator that had been given preference over MURA's machine, the question of location was thrown open to a nationwide competition. There is no available evidence that Lyndon Johnson played any personal role in matters related to the huge accelerator. But, from the Joint Committee to the AEC, and throughout the high energy community, it was recognized that, post-MURA, it would be politically infeasible to place the most expensive scientific installation in history—by 1967 cost estimates had risen from $240 million to $340 million—on either the East or West Coast. Within that tacitly acknowledged limitation, the high energy community was given carte blanche to be as precious as it chose in assuring that the taxpayers' $340-million investment would be conveniently situated for the handful of scientists that would use it. Thus, in announcing that it was open to site bids from throughout the nation, the AEC set forth criteria such as: "Proximity to a major airport having frequent service to major U. S. cities . . . to provide easy access and minimum travel time for university users and other visiting research personnel . . . Proximity to a cultural center that includes a large university [to] provide intellectual and cultural opportunities attractive for staff and families." [20] Some 125 sites, in 46 states, were presented for consideration. The AEC turned to the National Academy of Sciences for a politically germ-free appraisal, and an Academy committee, headed by the veteran troubleshooter, Piore (who had, at one point, untangled the Mohole Project for NSF) finally narrowed the list down to six—one on the East Coast, one on the West Coast, and four in the Midwest. Despite Johnson's announced policy of "developing and diffusing—throughout the nation—a strong and solid scientific capability," the chosen site was Weston, Illinois, within a few miles of the AEC's vast Argonne National Laboratory. A privately offered explanation for the site decision was that, since the political situation ruled out a site in the vicinity of the Lawrence Radiation Laboratory, Weston's proximity to O'Hare International Airport made it the "least inconvenient" of the available choices. As the most powerful participants in the politics of science, the physicists had been able to retain some degree of sovereignty as the new washed away the old. The biomedical scientists, however, were not so fortunate. In mid-July 1966, amidst reports that he had privately expressed concern about "too much research being done for the sake of research," Lyndon Johnson publicly stated, "A great deal of basic research has been done. I have been participating in the appropriations for years in this field. But I think the time has now come to zero in on the targets by trying to get our knowledge fully applied." [21]

Subsequently, Surgeon General William H. Stewart, though paying traditional homage to basic research, warned the biomedical research community, "Our society is fundamentally practical. It tends to be more interested in the product than in the process. I am confident that the public has accepted the importance of basic biomedical research. But the public expectations of research are framed largely in terms of specific medical benefit—the cancer cure, the wonder drug. It is important that we fulfill these expectations of tangible benefit whenever we can. . . . The gap, or the time-lag, between discovery and application is a source of increasing concern and criticism." [22]

And, finally, with no sugar coating at all, Ivan L. Bennett, former chief of pathology at Johns Hopkins University, speaking from his newly accepted position as No. 2 man in the White House science office, bluntly advised his colleagues to adapt themselves to the new politics of science. "Science . . . can no longer hope to exist, among all human enterprises, through some mystique, without constraints or scrutiny in terms of national goals, and isolated from the competition for allocation of resources which are finite. . . . Unless we biomedical scientists are prepared to examine our endeavors, our objectives, and our priorities, and to state our case, openly and clearly, the future will be difficult indeed." [23]

A quarter of a century had passed since Bush, Jewett, Compton, and Conant had led their profession into an alliance with the government of the United States. During those years, skilled leadership, good fortune, governmental understanding, as well as governmental indifference, had all contributed to the growth and productivity of American pure science. In the year 1967, the fourth year of Lyndon Johnson's Great Society, a new politics had enveloped the scientific community. Antiquated rhetoric still flowed from many of the statesmen of science, but a few saw through to the new realities of the science-government relationship: "The question is not whether we should have basic research," stated Donald F. Hornig, the presidential assistant for science and technology, "whether we should have research and development, or even whether it should continue to grow—but rather in what ways and for what purposes it should be expanded. The answer to this question will have to be supplied not by me but by all of us." [24]

At the conclusion of this inquiry into the politics of science, old and new, I would like to offer several personal observations.

The first is that, in assessing the claims, prophecies, and lamentations of the inhabitants of the scientific community, it is often useful to apply a generous discount. Their profession, having made a rapid ascent from deep poverty to great affluence, from academe's cloisters to Washington's high councils, still tends to be a bit excitable—not unlike a *nouveau riche* in a fluctuating market. The denial of a grant is equated with the persecution of Galileo; a harsh word from a know-nothing congressman, and the Cosmos Club simmers with talk of a new Dark Age.

Quite simply, I think it would be useful for the scientific community to calm down, and, if necessary, perhaps to impose its wondrously effective discipline on those who proclaim a state of crisis in basic research. Let it be recognized that, since World War II, each succeeding annual federal budget has surpassed its predecessor in funds for basic research; that no segment of American society has been accorded the degree of solicitude or wish fulfillment that the federal government has lavished upon pure science, and finally, that American science is the envy of the world, and properly so, for despite the lamentations that are today so prevalent, American science enjoys a combination of wealth and freedom unmatched in any nation. Because of competition for available public resources, American science cannot have everything that it desires. To depict this as a tragedy, when science is getting a great deal of what it desires, is to erode the credibility that is indispensable for the scientific community's dealings with the public at large and the United States Congress in particular.

Looking further into the science side of the science and government relationship, it would be desirable for the scientific community, on an institutional as well as on an individual basis, to pay more attention to its ethical atmosphere. I don't believe that a general fumigation is in order, but, from my own personal observations, and on the basis of what is commonly talked-of knowledge within the community, there is, to put it bluntly, a good deal of scientifically nonproductive chiseling that flourishes under the colors of scientific freedom. I would stress immediately that some distinction has to be drawn between this chiseling and what is referred to as "grantsmanship," i.e., the ability to draw money out of the public authorities. Grantsmanship, to my mind, is ethically neutral. Christopher Columbus, Joseph Henry, Charles Darwin, Samuel Morse, and Ernest O. Lawrence were grantsmen, and we are all better off for it. The issue is not the ability to get money; rather, it is the use to which the money is put. And, there is no doubt that because of the flood of federal funds that has come forth in the postwar period, large sums have come into the hands of people who are diluting, if not polluting, the atmosphere of the scientific community.

Thus, there is the phenomenon of some scientists flocking to a distant conference or laboratory for the ride rather than for science. Recently, the Office of Naval Research noted that "it might be safe to say that there is a correlation between the specific European laboratories visited by American scientists and the tourist attractions of the area in which they are located . . . Without question, the laboratories and work in many of the [European] metropolitan centers is superb and well worth a visit. At the same time, there is a surprising frequency of visits to laboratories where the work holds little relevance to the visitor's own professional interest and expertise." [25]

At times, costly equipment is purchased, not because there is a scientific purpose to which it is to be applied, but because money, usually federal money, is available. A scientist at a major medical school points to two electron microscopes, purchased, with federal funds, at a cost of approximately $30,000 each, standing crated in the basement of a laboratory building. "We have no use for them at present," he explained to me, "but we had the money and we figured we might as well get them." Scientists eager to enhance their reputations squeeze multiple publications out of one piece of research, thus broadcasting their names, but also cluttering up the literature. A young physicist frankly confides that he has published four papers on the results of a single experiment, when one, or perhaps two at the most, sufficed to convey his findings. "It is very difficult," he explained, "to make yourself known today."

The causes of conference-riding, needless purchase, and excessive publication are multiple and difficult to root out by edict. At the heart of the problem is the fact that, with the mechanization of much scientific research, it is now possible to function and thrive in scientific research without the sense of inspiration and commitment that characterized the community in its penurious days. Science was once a calling; today it is still a calling for many, but for many others, it is simply a living, and an especially comfortable one, for not only is it relatively affluent, but its traditions of freedom and independence provide a facade behind which government-subsidized liberties may be taken in the name of creativity. There was a time when many of these liberties, such as international travel, depended upon the personal possession of capital. Today, the significant factor is not one's own means; access to capital, not possession of it, is the hallmark of the man who has arrived in the New America.

The questions of whether these transgressions really matter, and, if so, what to do about them, are difficult. It is beyond the scope of this work to assess their effects on the quality and productivity of science, i.e., the substance of science. There is no doubt, however, that they have a heavily adverse effect on the politics of science. Having few peers in the misuse of public funds for personal purposes, congressmen are sensitively tuned to the misuse of such funds by others. To state it simply, a lot of science's money troubles grow out of the public's belief that science is careless with the taxpayers' money. Such were the lessons that Congress drew from the Mohole fiasco, from the rapid growth of NIH, and, to a lesser degree, from the long-running MURA episode.

As to what might be done to lessen public hostility, the goal is simple but the prescription is complicated.

The laissez-faire spirit and amorphousness of the scientific community provide a comfortable setting for those whose behavior elicits, and justifiably so, public concern for the uses to which tax funds are put. There is also no doubt that laissez-faire and amorphousness are beneficial to the progress of science. But there is no reason why the community itself, through its working institutions, professional societies, and publications, cannot elevate its own ethical climate. By and large, the peer system has worked well; in awarding funds and in judging material for publication, the peer system now might profitably place far greater stress on quality, rather than mere admissibility, on the advance of science rather than on the maintenance of the scientific community.

Finally, in looking at the science side of the science and government relationship, it would be very desirable for the community to encourage scholarly examination of just what it is that pure science contributes to society and what are the conditions that are most likely to foster valuable contributions. In response to congressional scrutiny and hostility, there has been an expansion of such inquiries. But, by and large, the statesmen of science are still telling Congress what they told it nearly a quarter of a century ago. In 1945, for example, David Griggs, chairman of the division of biology and agriculture of the National Research Council, advised a congressional committee "that it must be remembered that only one in a thousand basic researches ever pans out. But that one more than pays for all the rest." [26] Congress is not disposed to such long shots, and there is no reason why it should be. The odds on the utility of basic research, whatever they are, are better than a thousand to one; nevertheless, while congressional skepticism has mounted, many of the statesmen of science persist in ancient rhetoric.

On the government side of the science-government relationship, it would be desirable to introduce a little more coherence into government dealings with pure science. The anxieties and machinations of many members of the scientific community often are not without good reason. Research *is* a delicate process; productive research teams are difficult to assemble and easy to disrupt; research does not easily conform to the fiscal year. Though the relationship will never be tidy or friction-free, a good deal of trouble would be dispelled by the introduction of greater predictability and stability into government support of research. Thus, in the twenty-seventh year of the war-born partnership between science and government, it is time for Congress to make NSF what it was intended to be: the principal fount of government funds for the conduct of academic basic research and scientific training. There is no reason why other agencies cannot also continue to support these fields; experience suggests that diversity is useful, and, besides, the so-called mission-oriented agencies have particular needs to which they themselves can best attend in dealing with the universities. Nevertheless, there are too many strains, uncertainties, and irrationalities in the fictions that have been devised to finance basic research in this country. Though it runs against the congressional grain, and against general principles of executive responsibility, perhaps it is time to investigate carefully and sympathetically the possibility of putting basic research appropriations on a multi-year, perhaps five-year, basis. The disruptions that result from the annual vagaries of the congressional budget process are good neither for science nor for government. Both are quite mature now, and it is certainly time to put their partnership on a more trustful and stable basis.

Finally, of infinitely greater importance than the bothersome bookkeeping details that plague the relationship between science and government are the fundamental questions of purpose and value. What is it that man wants to do with his remarkable and growing power to investigate and manipulate his universe? In a world plagued by misery, is it decent for fine minds and great wealth to be dedicated to the interior of the atom and the mysteries of the planets? Or, as the ideologists of pure science would contend, does the unfettered spirit of inquiry provide the surest way to knowledge and salvation?

One month before his death, John F. Kennedy said, in an address to the centennial observation of the National Academy of Sciences, "Scientists alone can establish the objectives of their research, but society, in extending support to science, must take account of its own needs."

I think that statement is the supreme characterization of the predicament that underlies the politics of pure science.

NOTES

1. Don K. Price, "The Scientific Establishment," *Proceedings,* American Philosophical Society, Vol. 106, No. 3 (June 1962); also published in *Scientists and National Policy Making,* Robert Gilpin and Christopher Wright, eds. (New York: Columbia University Press, 1964).

2. "Science in U. S. Far Ahead, Says Top Russian," *Business Week*, December 24, 1966.

3. George A. W. Boehm, "The Pentagon and the Research Crisis," *Fortune*, February 1958, p. 134.

4. *Ibid.*

5. Nick A. Komons, *Science and the Air Force, A History of the Air Force Office of Scientific Research*, Historical Division, Office of Aerospace Research (Arlington, Va., 1966) p. 107.

6. *Federal Funds for Research, Development, and Other Scientific Activities*, NSF 65-13, p. ix.

7. *Senate Independent Offices Appropriations, Hearings*, 1963, p. 1347.

8. House Report No. 321, 87th Congress, 1961.

9. *Administration of Grants by the National Institutes of Health, Hearings*, Subcommittee of the Committee on Government Operations, 1961, p. 2.

10. *Ibid.*, 1962, p. 32.

11. *Ibid.*, pp. 20–21.

12. *Departments of Labor and Health, Education, and Welfare, Appropriations, Hearings for 1964*, Part 3, NIH, House of Representatives, p. 55.

13. Actions Taken by the Public Health Service to Strengthen Its Research and Training Grant Programs, *Departments of Labor and Health, Education, and Welfare, Appropriations, Hearings for 1964*, pp. 64–71.

14. Barry Commoner, "Government Research Grants: Effects of New Procedures on the Individual Investigator," letter to *Science*, June 7, 1963, p. 1048.

15. *Administration of Grants by the National Institutes of Health, Hearings*, 1962, p. 128.

16. Strategic Objectives for Training Activities, *Departments of Labor and Health, Education, and Welfare, Appropriations, Hearings for 1964*, p. 81.

17. *Ibid.*

18. Seminar on "Coordination of Federal Science Activities," July 18, 1966, unpublished transcript.

19. Statement of the President to the Cabinet on Strengthening the Academic Capability for Science Throughout the Nation, September 14, 1965, White House Press Release.

20. AEC Press Release, April 28, 1965.

21. White House Press Release, quoted in *Science*, July 8, 1966, p. 150.

22. William H. Stewart, "Biomedical Knowledge: Research, Development and Use; An Inventory of Opportunities and Cautions," address to Conference on Science in the Service of Man, Oklahoma City, October 1966.

23. Ivan Bennett, address to Conference on Science in the Service of Man, Oklahoma City, October 1966.

24. Donald F. Hornig, address to the American Physical Society, Washington, D.C., April 26, 1967.

25. *Science*, April 14, 1967, p. 228.

26. *Hearings on Science Legislation*, Subcommittee of the Senate Committee on Military Affairs, 1945, p. 602.

Afterword

I [dream] that I am Oliver Twist and I tremble as I shuffle down the aisle of the NSF orphanage, my empty soup bowl uplifted. The youngest orphans stare at me. They have no bowls, for they have never received NSF support. . . . Among the throng I spy the shocked and bewildered faces of scientific luminaries, prizewinners, even Nobel laureates. I raise my bowl to the Director and in a trembling voice ask, "Please, sir, one little laser."

"What," he thunders, "You have already had more than your share. *Out!*"

DANIEL KLEPPNER, Professor of Physics, MIT,
"Night Thoughts on the NSF,"
Physics Today, April 1990.

From the vivid expressions of hurt and the warnings of peril that continue to emanate from the science establishment, it is possible to conclude that the pre–World War II "orphan" described early in this book has never achieved full familial status in American society. The reality, of course, is that science has been deeply absorbed into the nation's economy and culture, and is indisputably regarded, perhaps uncritically, as essential to prosperity, national security, environmental protection, and fulfillment of our hopes for healthful longevity. It is accordingly supported by the federal government, industrial firms, private philanthropy, and the universities of this richest of nations, to the point where their combined expenditures for research and development far outdistance those of any other country. And it is defended and promoted by a strong network of lobbyists, ranging from "hired guns" in Washington that will take on any paying cause, to well-staffed scientific and professional societies that, along with individual university representatives, hover over science-related issues in the nation's capital. At a signal, they can inspire a blizzard of e-mails and faxes to any congressional office—and they have.

The National Science Foundation reported in 1997 that "the United States accounts for roughly 44 percent of the industrial world's R & D investment total and continues to outdistance, by more than 2 to 1, the total research in-

vestments made by Japan, the second largest performer."[1] U.S. R & D spending exceeds the combined sums of Japan, Germany, France, the U.K., Italy, and
Canada. (Russian science, in collapse for most of this decade, no longer enters
the list.) The proponents of spending even more for research point out that the
U.S. total is swollen by uniquely large spending on defense R & D. But deduct
the defense money and the American scientific enterprise still leads any other
nation in nonmilitary research by a wide margin. This is the case, too, for the
principal subject of this book, basic research in the U.S., for which government
remains the financial mainstay, though in recent years industry has substantially increased its expenditures in this area. From $2.4 billion in 1970, federal
support for basic science—performed mainly in universities, health centers,
and government laboratories—rose to over $18 billion in 1998; in the same pe-
riod, industrial spending for basic research grew from $530 million to over $8
billion. There is no calculating the "right" amount for research. But the numbers, and commensurate world-leading output of American science, establish
that the U.S. is not a laggard in devotion to science.

No matter. Even at this late date, the legitimate insecurities so evident in the
early post–World War II period remain prominent in the public rhetoric of science and, conveyed by a gullible press, in public perceptions of the state of science. (For example, a *Washington Post* article, March 5, 1997, headlined:
"Protesting Science's 'Slow Decay'—Coalition Urges Government to Boost
Investment in Research.")

A few elders of science have counseled against the incessant carping about
money, noting that science has generally evaded its political patrons' quite reasonable requests for priorities, i.e., a hierarchy of projects, from most important to least important. For instance, in 1991, while serving as president of
Rensselaer Polytechnic Institute, Roland W. Schmitt, a former GE research executive, cautioned that "it is hard to see that money alone is the problem."
Schmitt, a former chairman of the National Science Board, also observed, "We
are back to per capita expenditures near those of the 'golden era' of the late
1960s." True. Nonetheless, the lamentations of neglect have continued, as have
the prophecies of national harm if more money is not forthcoming.

Whether the fears are genuinely held, or are merely learned tactics that persist because they have worked in the past, is a separate question. Probably some
of both, but they fill the air. For example, Representative George E. Brown
Jr.(D-Calif.), who long ago embraced science as his holy congressional cause,
warned in 1997 that political support, hence financial support, for science is "a
mile wide and an inch deep."[2] His meaning—that many, especially in Congress, are for it but without strong conviction—would seem to be at odds with

the virtually intact survival of science spending on Capitol Hill through the worst of the budget-cutting "Republican Revolution" of 1995. In the Senate, the 1995 vows to reduce spending were quickly succeeded by Republican commitments to a rapid doubling of federal research spending, despite rigid limits on discretionary federal spending. Some senators proposed to do it over five years, others over ten or twelve years. None talked of reducing research spending. In his second term, President Clinton climbed aboard the research bandwagon, too, reversing the indifference that caused many scientific leaders to regard him as a disappointment in satisfying their special interest.

The enduring disparity between reality and rhetoric in the financial politics of the National Institutes of Health is illuminating. Between 1980 and 1990, appropriations for NIH rose in steady annual steps from $2 billion to $4.7 billion. The gain was "real." In purchasing power calculated at 1992 levels, NIH was $1.7 billion richer in 1990 than in 1980. As noted throughout this book, however, more is never enough in science. In 1990, Leon Rosenberg, dean of the Yale University School of Medicine, wrote a *New York Times* op-ed article lugubriously titled "Medical Research Is in Ruins."[3] Charging the government with fiscal neglect, Rosenberg stated, "Today our nation's health research program is burning, and the conflagration is spreading." Perhaps the alarming message worked. Starting with the $4.7 billion of 1990, Congress provided $5.5 billion for 1991; a rare dip, of $500 million, occurred in the following year, when antideficit politics began to take hold of federal spending. But after that, it was gain all the way, to 1999, when the NIH budget ascended to $15 billion.

In 1996, Nobelist Leon Lederman, still smarting from congressional termination in 1993 of the ultimate dream machine of high-energy physics, the 54-mile-around Superconducting Super Collider (SSC), declared that the nation is "riding a tide of anti-science." He did not mention that the termination of the SSC followed cost estimates that started at $4.1 billion in 1986, according to the General Accounting Office,[4] and eventually rose to merely vague assurances that perhaps $13 billion might do the job.[5] Lederman, a former president of the American Association for the Advancement of Science, later concluded that public support for science would be stimulated by a prime-time, major network television science series that would be the "*LA Law* and *NYPD Blue* of science and scientists." Though regularly professing inadequate funds for research, NSF and the Department of Energy provided $100,000 for a pilot script. The networks took a look, thanked Lederman, and no more has since been heard of that venture into stimulating public interest in science.

Searching for the origins of what it perceives as public indifference and political neglect, the science establishment insists that it suffers from a deficit of public understanding and affection. The head of communications for NIH, Anne Thomas, wrote in 1998—a year in which the NIH budget soared to a historic high—that public support is based "on a partial—and only partial—recognition of the link between research and better health. Consequently," she warned, "it may not be reliable over time."[6]

On the other hand, the National Science Board, on the basis of periodic opinion surveys conducted since the 1950s, reports the following: "Science and technology have become integral components of the American culture. Over 85 percent of Americans believe that the world is better off due to science, and this level of general support has continued over the last four decades. . . . Nearly 80 percent of Americans agree that the Federal Government should support basic scientific research that advances the frontiers of knowledge even when it does not provide any immediate benefits."[7]

From these observations, it may seem that the science and government relationship has undergone little change since the mid-1960s, when this account of the politics of science ends. My perception is that it has been stable in one important respect and turbulent in others.

The stability in government relations with science is in Washington's bureaucratic organization for the conduct and support of research. The structure has changed surprisingly little over these several decades, despite the growth of federal research and development spending from $15 billion in 1970 to $65 billion in 1998. Given the wealth and dynamism of the national research enterprise, it may be surprising to recall that the creation of NASA in 1958 nearly completed today's organizational structure for the administration of research by federal agencies. Several additions did occur through the amalgamation of small federal agencies, among them the creation in 1970 of the National Oceanographic and Atmospheric Administration and the Environmental Protection Agency. An especially horrifying change for the science establishment occurred in 1973, when President Nixon, distrustful of his science advisers, petulantly abolished the White House Office of Science and Technology—the scientific community's cherished connection to political power. Three years later, however, the connection was reestablished through a collaboration between the Ford administration and congressional friends of science. This time the White House office was rooted in an Act of Congress, rather than presidential preference, and its title was expanded to the Office of Science and Technology Policy. In substance, no change there.

The system's stability is also visible in the repeated failure of proposals to establish a cabinet-level federal Department of Science. The schemes have varied, but many of them would bring together the National Science Foundation, the basic-science programs of the Department of Energy, plus parts of NASA, the Environmental Protection Agency, and the National Institute of Standards and Technology. Proponents say cabinet status would bring visibility and more money for science. Opponents stress the value of diversity in the funding and administration of research. In any case, the department scheme comes and goes, without ever drawing substantial support.

Attempts have been made in Congress to rechristen the National Science Foundation as the National Science and Engineering (or Technology) Foundation to reflect the agency's expanded role in applied research. Invariably, these proposals have been easily defeated, upon the anxious insistence of the science clan, which fears the name change would reduce its prestige and budget share. During the Cold War, but especially after, many proposals were made to abolish or scale down the great laboratories born during or soon after the Manhattan Project and still operating under the Department of Energy, the descendant of the Manhattan Project via the Atomic Energy Commission. Staff reductions have been made, but the labs survive to this day, with an astonishing immunity to change. In 1997, the General Accounting Office reported that "although the nation has changed considerably during the decades since the War, DOE's contracting practices for its management and operating contracts have remained much the same as they were in the 1940s."[8] Plumbing the mystery of DOE's apparent indifference to post–Cold War reductions in its budget, the GAO found that "budget cuts did not result in commensurate reductions in spending. DOE spent almost the same in 1996 as it did in 1994. While Congressional funding decreased from $19.5 billion to $17.4 billion (or 11 percent) between those years," the GAO auditors reported, "spending only decreased from $20.4 billion to $19.9 billion (or slightly over 2 percent.)"[9] Explanation: DOE has a mountain of taxpayer money stashed away for uncertain times.

The U.S. commitment to a nuclear test ban once seemed to pose a bleak future for the Lawrence Livermore National Laboratory, one of DOE's two centers for nuclear weapons research. That is, until the Clinton administration designated the laboratory—located in vote-rich California—as the manager of nuclear Stockpile Stewardship, for which the principal instrument would be a newly constructed $1.5 billion laser device to test the ingredients of bombs without violating the ban on explosive tests. Military bases have finally been deprived of immortality by the imperatives of deficit reduction. Science establishments have taken their place.

Against this background of institutional stability, turbulence comes principally from two directions: ideological contention over the appropriate role of government in industrial research support, and the competitive strivings for survival within a mature, sometimes ossified, research establishment that has had to make its way through deficit-reduction politics.

Ideological strife in research policy arose when the Clinton administration sought to expand commercially relevant research programs in the Department of Commerce and in the Pentagon. In an odd turnabout, Republicans, traditionally the party of big business, denounced these efforts as "corporate welfare," and choked their growth or killed them outright. Republicans are now the champions of basic research in universities—and let industry fend for itself; Democrats are also for basic research, but, true to their interventionist tradition, they aspire to a bigger government role in research that is directly, rather than distantly, relevant to industry.

Scientific strongholds, in both government and academe, face the eternal dangers of aging institutions: changes in the circumstances and priorities of their patrons, accelerated by the end of the Cold War, and competition from aspirants seeking to enlarge their gains or simply join the game. Benefiting from long-established political connections, the *ancien régime* that arose during World War II has resisted dethronement, usually with success. As noted, the DOE bomb laboratories have long survived their original mission. And proposals to bring competition into the award of agricultural research funds—à la the NIH and NSF models—have made only token headway against the old-time beneficiaries in the land grant colleges. But the bite of financial need under constrained budgets has produced unprecedented carping over who gets the money. Public protests from scientists outside high-energy physics, including some physicists, helped bring down the Superconducting Super Collider. And the managers of the Space Station are plagued by scientists who get up and say, in effect, it may be a great project, but it ain't science.

When it was first published in 1967, *The Politics of Pure Science* received favorable reviews—from outside the scientific community. For example, the leading sociologist of science, Professor Robert K. Merton of Columbia University, described it as "a book of consequence . . . that could be understood and should be read by the President, legislators, scientists and the rest of us ordinary folk."[10] Gratifying words, of course. But also of interest were the responses from *inside* the scientific community, for they provided additional instruction in the *realpolitik* of science. From inside came a flock of grumpy reviews by elders of science and their acolytes in publications serving the scientific community. One of these reviews was preceded by a touching attempt to soften the tale of *The Politics of Pure Science.*

Informing me that he had been commissioned as a reviewer by *Scientific*

American, Victor Weisskopf, a revered figure of the physics establishment, asked me to meet him in Cambridge, where he was recovering from a skiing injury. When I arrived, he waved a sheaf of proofs and, with the assertiveness of a prof turning back a deficient term paper, declared that changes would be necessary. If not carried out, he intimated, his *Scientific American* review would reflect that failure. Had he found errors? I asked. No, he acknowledged. Rather, he wanted the work "toned down" because "it makes us look bad." Science is a politically delicate enterprise, he explained, and might be harmed by the book's description of its inner workings. At that time, with six or seven years of experience in reporting on the scientific community, I was familiar with what I faced: the belief that anything but reverence for science and its practitioners constituted hostility. The conversation essentially ended there.

The book was published as written, and the Weisskopf review assailed it, as he indicated it would.[11] Stating that "Greenberg's book . . . emphasizes the frailties of the scientific community," he added, "It is not, however, much counterbalanced: positive achievements are conceded only grudgingly. . . . As he describes the financial support of science today," the review continued, "he loves to tell us about the quarrels, the ambitions and the weaknesses of scientists. He is most eloquent when he has dug out some unhappy tale"—and so on for several reiterations of those themes. To which I plead, Yes, but.

I have no quibble with Weisskopf's assertion that "Greenberg regards [the scientific] community with the air of a man who has studied a strange tribe that is quite different from the rest of humanity, and he finds the members of the tribe quite impressed with themselves." True, then and even to this day.

Then there was the review by a tin-eared minor official in the federal science bureaucracy, uncharmed by the alliteration in the title *The Politics of Pure Science,* in which he detected "derisive overtones."[12] "Reference to 'pure science' rather than 'basic' or 'fundamental' or even 'undirected' science," the reviewer complained, "tends to imply that research arising from practical motivations is somehow 'impure.'" To which I plead, Nonsense, suspecting that for the reviewer in question, the title would have passed without notice if it had been: *A Preliminary Examination of Selected Interactions between Academic Researchers and Institutions and Federal Agencies in the Prioritizing, Selection, and Financial Support of Scientific Inquiry* (with apologies in advance, since such a title probably exists in the voluminous literature of science and government relations).

Despite the carping reviews, or possibly because of them, *The Politics of Pure Science* sold moderately well, considering that the title is a puzzler outside the scientific community and that there is little general interest in the obscurities of science and government relations. It enjoyed several printings in a clothbound edition and a life of more than ten years in paperback. By the early 1980s, however, it was out of print, though enough copies existed to satisfy many dedicated book hunters. But not enough copies, finally, as the years went by.

This resurrected edition originated in the small but steady stream of inquiries for *The Politics of Pure Science,* including some from university professors who read it in their student days and now wanted to use it in their own courses. When the University of Chicago Press and I first discussed the project, we thought in terms of an update that would retain the original text and bring the story up to the present time. But then we agreed that too much has happened to meld the old and the new into a manageable account. The decision went to republishing the original text, unchanged, except for the introductions by John Maddox and Steven Shapin and this afterword. The events that followed the writing of *The Politics of Pure Science* will be related in a new book, now in progress, also to be published by the University of Chicago Press.

D.S.G., WASHINGTON, OCTOBER 1998

NOTES

1. *Science and Engineering Indicators, 1998,* U.S. National Science Board.
2. *Science and Government Report,* February 1, 1996.
3. *New York Times,* September 2, 1990.
4. *Nuclear Science, Information on DOE Accelerators Should Be Better Disclosed in the Budget,* GAO/RCED-86-79.
5. *New York Times,* October 26, 1993.
6. "Public Support for Medical Research—How Deep, How Enduring?" *Academic Medicine,* journal of the Association of American Medical Colleges, February 1998.
7. *Science and Engineering Indicators, 1998.*
8. *Department of Energy: Contract Reform Is Progressing, But Full Implementation Will Take Years,* GAO/RCED-97-18.
9. *Department of Energy: Funding and Workforce Reduced, but Spending Remains Stable,* GAO/RCED-97-96.
10. *New York Times Book Review,* October 6, 1968.
11. *Scientific American,* March 1968.
12. Joel A. Snow, *Bulletin of the Atomic Scientists,* May 1968.

Index